ROBOTS ARE
PEOPLE TOO

ROBOTS ARE PEOPLE TOO

How Siri, Google Car, and
Artificial Intelligence Will Force
Us to Change Our Laws

JOHN FRANK WEAVER

 PRAEGER

AN IMPRINT OF ABC-CLIO, LLC
Santa Barbara, California • Denver, Colorado • Oxford, England

Library of Congress Cataloging-in-Publication Data

Weaver, John Frank.
    Robots are people too : how Siri, Google Car, and artificial intelligence will force us to change our laws / John Frank Weaver.
       pages cm
    Includes bibliographical references and index.
    ISBN 978–1–4408–2945–1 (hardback) — ISBN 978–1–4408–2946–8 (ebook)
1. Artificial intelligence—Social aspects. 2. Artificial intelligence—Law and legislation. I. Title.
Q335.W386   2014
343.09′99—dc23   2013023469

ISBN: 978–1–4408–2945–1
EISBN: 978–1–4408–2946–8

18 17 16 15 14   2 3 4 5

This book is also available on the World Wide Web as an eBook.
Visit www.abc-clio.com for details.

Praeger
An Imprint of ABC-CLIO, LLC

ABC-CLIO, LLC
130 Cremona Drive, P.O. Box 1911
Santa Barbara, California 93116-1911

This book is printed on acid-free paper (∞)

Manufactured in the United States of America

Portions of Chapter 1 were adapted from John F. Weaver, "Siri Is My Client: A First Look at Artificial Intelligence and Legal Issues," *New Hampshire Bar Journal* 52, no. 4 (Winter 2012), with permission from the New Hampshire Bar Association.

For Alicia and Ella, and Baby Boy Weaver,
who I hope will find me a nice robo-nurse in 60 years.

Contents

Acknowledgments

Many thanks to Beth Ptalis for guiding me through the proposal, writing, and publication processes. I would also like to thank David Cope, Elizabeth Kazakoff, and Peter Kazanzides for the time they took to discuss their work and opinions with me.

I am deeply in the debt of Jen Finch, who was more helpful and cooperative than any rational human being should have been. She was the research librarian who heard "Can you find me articles about robots performing surgery without a human being doing anything?" and said "No problem" rather than "No way."

Finally, I have endless gratitude for my wife, best friend, and favorite editor, Alicia Weaver, who read every word I wrote and properly told me when I should erase words and replace them with better ones.

Part I

MAY A ROBOT HURT A HUMAN BEING?

CHAPTER 1

It's Not What Isaac Asimov Promised, but Artificial Intelligence Is Here

To be clear, this is not a book about HAL, KITT, Data, or any of the droids from *Star Wars*. This is not a book about machines smarter than people or software that matches human intelligence. This is, rather, a book about machines that recreate *some aspect* of human intelligence or are capable of making a limited type of human decisions. That is to say, this is a book about artificial intelligence (AI) that is already available or soon to be available. And make no mistake about it, Apple's Siri is just as much AI as C-3PO.

The difference between the AI in C-3PO and the AI in Siri (besides the voice of Anthony Daniels) is the difference between strong and weak AI. Strong AI matches or exceeds human intelligence and can therefore solve any problem and interact in any social situation much like a person would. Strong AI is pure fiction at this point. However, weak AI is another matter altogether. Weak AI recreates only elements of human intelligence in a computer. We interact with weak AI all the time—Google's search engine, Global Positioning System (GPS), video games, etc. Any machine or software that is able to follow simple rules to replicate human intelligence qualifies.

What we have seen recently is a rapid advance in weak AI. Famous examples of this development include Deep Blue, the chess master machine, and Watson, the Jeopardy master machine. But in a sense, those examples are just as fantastic as anything in *Star Wars* or *Star Trek*. Hardly anyone interacts with machines like that.

We are about to see an explosion in weak AI products that are commercially available. The products listed above exist in limited functions: Google is only available on the Internet; you don't *need* GPS to travel anywhere; video game AI only functions within a fictional world. In the next 10 to 20 years, AI will permeate our lives more thoroughly, from the way we travel to the media we consume to the tools in place to police our streets. As weak AI becomes an increasingly inescapable presence, it will force us to change some of our assumptions about the world we live in.

This includes assumptions made by our laws. In America, almost all of our laws at the local, state and federal levels share an assumption: all decisions are made by human beings. The development and proliferation of AI will force changes to our laws because many of them will not adequately address how AI products interact with us and each other. Areas of law as diverse as liability, intellectual property, constitutional rights, international law, zoning regulations, and many others will have to adjust to effectively account for decisions that are not made by human beings.

Having said all that, a quick look at strong AI in fiction can provide a better idea of what weak AI is and what this book will address.

The Three Laws of Robots and the Personhood of AI

One of the seminal ideas about AI in science fiction is the concept of the Three Laws of Robots. In 1942, science-fiction writer Isaac Asimov introduced this idea in his short story "Runaround." Although they are frequently tweaked by other writers, the original formulation looked like this:

1. A robot may not injure a human being or, through inaction, allow a human being to come to harm.
2. A robot must obey the orders given to it by human beings, except where such orders would conflict with the First Law.
3. A robot must protect its own existence as long as such protection does not conflict with the First or Second Laws.

If you look closely at the three sections of this book, you'll find they are based on these laws. The Three Laws of Robots try to ensure that robots will work for the betterment, and not to the detriment, of human beings. Of course, if that worked out well, there would be little conflict and little of interest in a lot of science fiction. Rather, as Roger Clarke has noted, Asimov's stories themselves provide ample evidence that there is no reliable way to constrain the behavior of robots through a set of rules.[1]

This creates an inherent conflict for AI robots and software in fiction that is explored frequently: strong AI has the same free will as any human being but does not have the freedom of action given to actual human beings. Asimov's "Runaround" features a three-person crew traveling to Mercury. One of the crew members is a robot. When a dangerous trip to obtain a rare element is necessary, the robot is ordered to undertake it. He has free will, but it is conditional. Is that really free will?

Even science fiction that does not address the Three Laws addresses the potential for strong AI to have free will. In *Star Trek: The Next Generation*, the android Data is forced to defend his sentience and personal freedom in a military courtroom. In the film *2001: A Space Odyssey*, the question of whether the HAL 9000 has emotions and free will is crucial to the plot. In Ray Bradbury's "Marionettes, Inc.," from *The Illustrated Man*, one of the main characters purchases an AI duplicate of himself, intending that the robot will take over for him in his home for periods of time, so that his wife will not miss him. Although the manufacturer guarantees that the robot will obey orders—"No strings attached" is its motto— the AI duplicate turns on him to begin a relationship with the man's wife. In story after story after story in the genre the message becomes clear: partial free will and recognition as a person are insufficient for strong AI.

Again, to be clear: none of the weak AI that is discussed in this book, that is available now and that will be available soon, raise questions of free will and recognition as a full person. The machines and programs re-create only some aspect of human intelligence or decision making. But even in recreating only an aspect of human decision making, weak AI makes decisions we assume are made by people. Under the law, the "technical legal meaning of a 'person' is a subject of legal rights and duties."[2] As our laws are currently constituted, machines do not have legal rights and responsibilities; the people who design, manufacture, sell, and use them do. What happens when weak AI—through no fault of the people who have designed, manufactured, sold, or used it—causes damages? Causes death? Creates valuable intellectual property? Violates someone's constitutional rights?

Even when the AI is not capable of free will, the law should recognize that limited rights and responsibilities are awarded to weak AI in order to protect the actual people who interact with them and, ideally, to put humanity in a better position to benefit from it. How the law should change to recognize AI robots and programs as limited "persons" in this way is the focus of this book.

Siri—Apple's Prologue to the Coming AI Revolution[3]

Although weak AI products have been commercially available for some time, the first mass-marketed product that truly provides a glimpse of the capabilities and legal issues inherent in forthcoming AI is Siri, the "intelligent personal assistant that helps you get things done just by asking."[4] Siri is the first commercially available, advanced weak AI. By "advanced weak AI," I mean weak AI that can recreate elements of human intelligence through humanlike interaction.[5] This distinguishes it from other popular examples of weak AI like Google—in their natural setting, humans don't type questions and requests to each other.[6]

Why is advanced weak AI important? Clifford Nass, a professor at Stanford University and the director of the Communication between Humans and Interactive Media Lab, believes that when we talk to a computer and it answers back, we treat it like a real person. "The human brain is built for speech, so anything that sounds like a voice, our brains just light up and we get an enormous range of social and other responses," he says. "Our brains are built to treat these conversations with computer-based voices to an incredible degree like [conversations] we are having with actual people—including flattery, flirtation and all the rest."[7] Despite ourselves, we will treat advanced weak AI like the other people in our lives, meaning the way we trust and incorporate advanced weak AI into our lives will be different from how we treat machines now. Despite its inconsistent performance—and its possible inferiority to Google Voice Search[8]—Siri represents other forms of forthcoming AI that we will interact with as human beings, raising the question of whether we should revise our laws to reflect the real way we actually treat these machines.

Like the Internet and GPS, Siri is the product of DARPA, the Defense Advanced Research Projects Agency, a Defense Department agency that develops new technologies. In 2003, DARPA entered into an agreement with SRI International, a research group in Menlo Park California, to research and develop a "cognitive assistant that learns and organizes."[9] In 2007, SRI incorporated a separate entity, Siri, Inc., to develop a commercial application for the AI developed in the DARPA project.[10] Siri, Inc. received venture capital backing and eventually released a virtual personal assistant application for the iPhone in February 2010.[11] In April 2010, Apple, Inc. bought Siri, Inc.[12] When Apple released the iPhone 4S in October 2011, Siri was preinstalled as an all-purpose virtual assistant, not merely as an app.

Siri is not just a virtual personal assistant. It has both speech input and speech output. Users can speak to it and receive spoken responses. The degree of its interaction with the iPhone and the Internet is fairly extensive. Ask Siri about the weather, and Siri will give you a short summary of the weather forecast where you're located. Ask Siri to tell your husband you are running late, and Siri will send a text message. Ask Siri to schedule lunch with your mom next Friday, and Siri will update your calendar and give you verbal confirmation. Siri does not process speech input solely on your phone. Rather, the software sends commands through a remote server, so it is necessary for users to be connected to Wi-Fi or carrier service (Verizon, AT&T, etc.).[13]

The actual process Siri uses to translate your words into action—"breath to bytes"—represents an impressive achievement of Internet technology. Let's say you want to tell Siri to text your best friend, Lando. Siri would immediately encode the sounds of your speech into a compact digital form that preserves its information. Siri would then send that digital bundle of information wirelessly through a nearby cell tower and through a series of fiber optic cables or other landlines back to your Internet Service Provider where it could communicate with a server in the cloud. A series of data-based models, designed to comprehend spoken language, would analyze your speech.

At the same time, software within Siri itself would evaluate your speech. Working with the server in the cloud, Siri determines whether the command can be best handled on the phone—like if you asked it for a picture of Lando on the *Millennium Falcon* that was downloaded—or if it must connect to the network for further assistance. If the server is needed, it will compare your speech with a data-based model to estimate what letters might constitute it. The server then uses the highest-probability estimate to proceed.

Based on this estimate, your speech—now understood as a series of vowels and consonants—is analyzed using a language model, which estimates the words that your speech contains. The computer then creates a list of likely interpretations for what your speech might mean and chooses the most probable. If Siri and the server have worked properly, then they have determined that your intent is to send a text, recognized that Lando is your addressee, pulled his contact information from your phone's contact list, and properly drafted your actual note to him. But all you see is your text message, which magically appears on-screen, no hands necessary. If, however, Siri becomes uncertain during that analytical process, the computers will defer to you with further questions: did you mean Lando Calrissian or Lando Lakes, etc.[14]

In short, Siri depends on sophisticated weak AI in the phone software interfacing with sophisticated weak AI in the Internet cloud to create a user-friendly advanced weak AI program. By relying on central processors over the Internet, Siri is able to improve as more people use it. Siri collects data from users and analyzes that data to improve its services.[15] That represents a lot of autonomous technology interacting with real-world people, which can result in legal issues we haven't seen before, but that will become more common as the AI products discussed in this book come to market and are widely used. In this regard, Siri can be seen as the first shot across the bow of the current legal assumption that almost all decisions are made by human beings. By following Siri's functions to a logical extreme, we can preview future AI cannonballs that will sink that assumption. In particular, Siri demonstrates that intellectual property (discussed further in chapter 8) and liability laws (discussed further in chapter 2) will have to adjust to accommodate AI.

The Sounds of Siri—Who Owns the Copyright?

The words, sentences, and ideas expressed by Siri have little value as recorded media—for now. But Siri poses the potential to contribute to successful and profitable media, and once there's real money involved, the question of who owns the copyright to a Siri response, and potential royalties related to its commercial use, becomes more important.

Sampling with Siri

Sampling in various forms of popular music—most notably hip-hop, pop, and rap—has been a hotly contested issue in copyright law. As the Sixth Circuit noted, "Advances in technology coupled with the advent of the popularity of hip-hop or rap music have made instances of digital sampling extremely common and have spawned a plethora of copyright disputes and litigation."[16] Copyright disputes involving sampling have resulted in litigation concerning artists as diverse as Roy Orbison,[17] the Beastie Boys,[18] and George Clinton Jr.[19] George Clinton's music, in fact, is responsible for one of the largest allegations of copyright infringement, as entities holding copyrights to his music filed 500 counts against approximately 800 defendants for using portions of his recordings and music without authorization in 2001.[20] In these cases, the key dispute was between the owner of a musical composition or sound recording and another person who sampled that music or sound to create otherwise new media.[21]

With Siri, a user could experiment with Siri's speech output in order to create an amusing or interesting sound bite, record that sound, and alter it into commercially successful media (e.g., song, commercial, etc.) using Auto-Tune. Under that scenario, it appears that this creator would own the copyright to the new media. However, the law is not established on this point.

Ownership of Intellectual Property Produced with Siri

Section 201(a) of the Copyright Act states that "Copyright in a work protected under this title vests initially in the author or authors of the work."[22] The Supreme Court has stated that "As a general rule, the author is the party who actually creates the work, that is, *the person who translates an idea into a fixed, tangible expression entitled to copyright protection*"[23] (emphasis added). But this idea is in flux as an increasing number of weak AI computer programs like Siri are able to produce original music, text, and sound.[24] You can find books of computer-generated poetry,[25] novels written by a hacked Macintosh that draws inspiration from Jacqueline Susann,[26] and CDs of music composed by an autonomous program.[27] When Siri says something new, who is the "person" who owns that copyright?

Ralph D. Clifford[28] attempted to answer this question, albeit while examining intellectual property created through human-computer partnerships rather than intellectual property created by Siri specifically.[29] Clifford examines the potential for computers and machines to develop copyrights, deciding that there is a spectrum of human-machine interaction. On one end of the spectrum, there is intellectual property developed by machines that are programmed and guided by humans; copyrights from those pairings are properly owned by the programmers.[30] On the other end of the spectrum are machines that are capable through their programming of creating new media with little to no further interaction with people; copyrights from those machines enter into public domain, under Clifford's analysis.[31] He relies on the Supreme Court's reasoning from *Feist Publications, Inc. v. Rural Telephone Service, Co.*: "The *sine qua non* of copyright is originality. To qualify for copyright protection, a work must be original to the author."[32]

Clifford extends this reasoning, explaining that a machine that is programmed to create creativity on its own does not warrant copyright protection because "no one derives rules for the computer to control its creativity; rather, using its learning algorithm and based on the training examples it is given, it develops rules on its own. This learning is done independently of its user."[33] And in fact, this sounds strikingly similar to Siri's ability to learn to operate better as more people use it. As Clifford asks, "[W]ho then can claim a copyright in the expressive works" created solely by a computer?[34] He answers his own question: "The claim of the user of the machine seems highly dubious. The user was not the originator of these expressions as no specific creative effort was exerted by the user."[35]

I suspect that Clifford is mistaken regarding the intellectual property developed by programs like Siri. Nature and the law abhor a vacuum, which is effectively what the public domain is for intellectual property. One of the reasons why Siri is so interesting is that it is the first *commercially available* advanced weak AI. Although Siri is likely on the end of Clifford's spectrum where intellectual property is owned by the human interacting with the AI, that spectrum breaks down when the AI is available to a mass market. Clifford assumes that the programmer (the person who creates the machine), the user (the person who uses the machine to create output), and the artist (the person that makes new media with the output of the machine) are all the same person. Beginning with Siri, that is not necessarily the case.

Rather, if there is an owner to the sounds Siri creates, it is likely either one of or a combination of the following parties:

- The artist who creates the new media using the sample of Siri's speech output;
- The user who experimented with Siri in order to create the speech output (this could be the same person as the artist); or
- Apple, which owns the Siri source code.

Thus far, there has been little if any difference between these three parties when other forms of weak AI have created new copyrights. When Scott French hacked his Macintosh to create prose reminiscent of *Valley of the Dolls*, he programmed the computer to ask him questions that would permit it to draft appropriate language and story. When David Cope entered into an agreement with Centaur Records to release a CD of music composed by the program he named "Emily Howell," he had written the Emily Howell program and provided it with musical inputs to analyze before it wrote new music.

Cope is an apt example because his Emily Howell program is the direction that weak AI is going. Emily Howell is a program that can analyze music and feedback and create its own style.[36] Cope says, "I've taught the program what my musical tastes are, but it's not music in the style of any of the styles—it's Emily's own style."[37] With regard to other human composers, Cope says that they "are looking at a competitor—a virtual composer competing in the same arena with 'her' own style and music that is really excellent. It seems to me that these composers should feel a little less smug and more defensive about their position."[38] Although he did not make any public comments regarding the copyright of Emily's first CD when it was published by Centaur Records in 2010,

Cope has since said that he believes he owns the copyright to any music Emily produces, and it can be safely assumed that he kept any royalties as Emily's programmer.

Who Should Own Siri-Created Media and Similar Copyrights?

Cope and Emily demonstrate why *someone* has to own the copyright associated with media produced by Siri, and other similar computers, machines, and programs. When there is money involved, the public domain is not a realistic option, as someone (likely many people) will claim ownership and do so with compelling arguments. Similarly, without the potential for return on an investment, programmers and investors will not seek to devise technologies that develop creativity.

Although no law speaks directly to this issue now, as weak AI is more commercially available, Congress or a court will need to address it in the near future. This necessary development will be discussed more thoroughly in chapter 8, but for now, let me say that from a public policy standpoint, it would be better for Congress to address it, so there is unanimity throughout the country, without divisions between circuit courts. Assuming that there will be more mass-marketed sophisticated weak AI like Siri, granting ownership of media created with mass-market AI to the users and artists rather than the programmers will stimulate greater creative development of media using that AI. Additionally, the users and artists are closer to the "author" of such media as currently contemplated under U.S. copyright law.

But Clifford's concern is a legitimate one. Should traditional copyright and patent protection be granted to the writer of code that produces creative products largely independent of human direction? Emily Howell and its predecessor program Emmy (which was less advanced musically but more autonomous as a composer) already produce classical music; there is no reason to believe a program that produces commercially successful pop music will not be developed sometime in the near future. Similarly, there are AI programs that write full articles regarding sporting events;[39] there is no reason to believe a program that produces commercially successful novels will not be developed in the not-so-distant future. One of the suggestions discussed later in this book is that there should be a line drawn between AI that creates new content or media completely autonomously and AI that requires human interaction to create. In recognition of the concerns cited by Clifford for properly fostering and rewarding human creativity, the copyrights owned by the programmers of AI that can create completely autonomously could receive a limited copyright—maybe 10 years—before it enters into the public domain.[40]

Although this issue seems remote now, Siri is only the first product in an increasingly complex line of weak AI products that will make issues of intellectual property ownership more important, as chapter 8 explains.

Siri Made Me Do It—Liability for Siri's Mistakes

There are already instances in which accident participants have blamed weak AI—most notably GPS—for the accident. For example:

- In 2009, a driver followed the directions provided by his GPS, until he was trapped on a narrow cliffside path—described by a local farmer as "one he would not even take his horses down"—and the police had to tow him back to the main road. Despite that driver's attempts to blame the GPS system, the British court found him guilty of careless driving.[41]
- In 2012, a driver drove through a "No Entry" sign onto a road, pulling into the path of another vehicle. Although they were obscured by vegetation, there were similar signs for 180 yards leading up to the intersection. The judge found the driver's claim that he was following directions from his GPS unconvincing and fined him.[42]
- In 2011, a judge found that a GPS device was partially to blame for a fatal car crash, stating that a driver placed too much reliance on the GPS's map of the road he was on. The judge sentenced him to over four years in prison.[43]

Although there is limited case law addressing GPS culpability in accidents, it is not far-fetched to suggest that a court could attribute partial or total liability to a GPS system in some instances. Some American attorneys expect that development sometime in the near future as more drivers depend on satellite directions, particularly in areas where they do not know the geography.[44] That dependence will increase as advanced weak AI in GPS improves to the level suggested by Siri. As drivers are able to talk directly to their GPS—"There's a traffic jam up ahead—is there another way around the next few intersections?"—they will rely on those devices more and more. By mimicking actual human interaction, GPS devices increase the likelihood that a court will attribute partial or total liability to a GPS system because it becomes more reasonable for drivers to rely on them; if it talks like a human, we treat it and trust it like a human.

Siri invites a similar reliance—it is programmed to accumulate user data through the Internet cloud where it operates, learning more about users as they ask questions and make requests. By design, Siri tries to make itself more indispensable with each user interaction.

So consider this scenario: A user is visiting a city that he has not traveled to before. He has an appointment, he is lost, and he asks Siri the fastest route to the site of his meeting. Siri provides him with directions, but does not mention some of the potential dangers associated with that route. If the route takes the user through an area with a high crime rate, and he is assaulted, mugged, etc., what liability does Siri (and by extension, Apple) have for those damages?

This is not an easy question to answer, and any decisions would depend on a court's analysis of the user's actions. Did the user behave reasonably? Should the user have identified the dangerous situation? Did the user have sufficient information to deviate from Siri's directions? Apple could argue that the user was contributorily negligent in causing an accident if he failed to act reasonably or exposed himself to unreasonable risk of harm.[45]

Any court considering the question of reasonableness in the scenario described above will have to consider Siri's design. It is programmed to become more useful. With that goal in mind, isn't it reasonable for users to increase their reliance on Siri as they use it more? This is particularly true where users are asking questions concerning topics of which they know little—directions in a new geographic location, recipes for cooking, lesson in carpentry, etc. The interface matters in this analysis, as well. It is more reasonable to rely on a device that responds to conversational questions than on one that requires a cumbersome keyboard.

There is no quick answer to this, but Siri is only the first of many new technological developments that will require a reexamination of how we determine and assign liability. For example, numerous American companies—including General Motors and Google—have spent considerable time and resources developing an autonomous car.[46] Current prototypes are already traveling—under supervision—along highways at 70 miles per hour and through well-populated downtown areas in selected states; Alan Taub of General Motors predicts that autonomous cars will be commercially available by 2020.[47] While tort law has assumed the validity of driver negligence since the invention of the automobile, that concept could soon be outdated, like elevator operator negligence. Just how outdated will be explained in chapters 2 and 3, which examine AI liability and autonomous cars, respectively.

What's on the Horizon after Siri

I hope that this brief discussion about the issues Siri could introduce into courts, contracts and legislatures demonstrates how our laws are not designed to contemplate AI in any form. And, quite frankly, they shouldn't be: Henry Clay, Daniel Webster, and Jefferson Smith never had any reason to consider machines that could think.

But we do now. There are products discussed in this book that are on the market now, that are coming to stores in the not-so-distant future, and that are in development. They include autonomous cars (chapter 3), robot babysitters (chapter 4), attack drones (chapter 7), writing and composing software (chapter 8), AI farm and factory workers (chapter 5), and police surveillance robots (chapter 6). It is possible that weak AI will be a revolution in our lives similar to the Industrial Revolution in the 19th century, the spread of automation in the late 20th century, and the rise of the Internet in the 1990s and 2000s.

That is to say that people 30 years from now will wonder how we ever got along without AI, but also that AI will destroy a lot of jobs and disrupt a lot of lives in the next few decades. With that in mind, it's important that we start having a conversation now about AI, public policy, and our laws, particularly because the timetable that permitted a gradual adaptation to the Industrial Revolution no longer exists. Depending on the source, the Industrial Revolution in America started somewhere between the 1790s and 1830s. However, it took over a century for the country to adapt its laws to address the issues it created: wealth that wasn't land based; large, concentrated labor forces; work that did not center around the family or home; new workplace hazards; new environmental hazards; etc. The U.S. Congress and state governments did not enact child labor laws, minimum wage laws, public education reform, collective bargaining rights, and other legal innovations until well into the 20th century. And it was not until the mid-20th century that the wealth generated by the technological advances of the Industrial Revolution spread widely enough to create a large and secure middle class based on manufacturing jobs. That wealth would not have spread without the laws passed in response to the Industrial Revolution, which chapter 3 explains.

Unfortunately, we don't get a century anymore to respond to technological advances. Based on the timing of the last two major technological developments—industrial automation and the Internet—we get only 20, maybe 30, years to ensure that the benefits of technological advances are widely spread. To protect the middle class, we'll have to quickly modify our laws to adapt to technological changes. So we need to start talking about AI and the law now.

The rest of this book strives to launch that conversation by discussing public law (law that is publicly debated and promulgated: legislation, regulations, etc.) and private law (law that exists between individual people and companies, like contracts, torts, etc.) that should change to properly govern AI. Although you have to follow Siri to an extreme use of its programming to reach the challenges discussed above, the other programs, robots, and products discussed here don't require such an exercise. They will introduce those challenges by their very presence on the market or by their governmental use. In many cases, laws will have to treat AI as legal "people," assigning them certain rights and responsibilities to improve the lives of actual people.

Before moving on, it's helpful to look at some questions and consider both whether there are correct answers right now and what you think of the technology that makes these questions necessary:

- If two autonomous cars are in an accident, and the cars did not cause the accident but rather acted in the best possible way to avoid injury, is someone hurt in that accident barred from recovering from the owners of the cars or their insurance companies?
- Can an AI surgeon commit medical malpractice?
- What level of weak AI should be regulated by government agencies? Should Siri? Should autonomous cars? Should writing and composing software? Should spy drones?
- As most city and town ordinances are currently written, could a robot qualify as "adult entertainment"?
- If you place your children in the care of an AI babysitter, are you liable for everything that goes wrong? Is the babysitter's manufacturer?
- What constitutional limits exist to control police use of surveillance drones?
- Could other countries use AI drones in the United States without it being an act of war or aggression?
- If a programmer creates software that can produce books as popular as *Harry Potter* every week, should she keep the billions of dollars the "author" earns?
- If we know so many AI products are coming, are there public policies that will help us mitigate job losses?

The upcoming chapters will address these questions and many others.

CHAPTER 2

How to Sue a Robot: Liability and AI

Some forms of AI are so common that people don't even consider them AI—Google, Siri, videogames, etc. Most people aren't typically worried about liability when I talk about those AI. But when the topic changes to exciting forms of AI that are forthcoming—Google Car! Autonomous surgeons! Robot nannies!—liability *immediately* becomes a big concern. "Who would want a car that drives itself?! What if you hit someone?" "No one will make a robot babysitter—think about all the lawsuits!" "Don't lawyers cause enough problems for doctors? You think they're going to open themselves up to those lawsuits?"

These concerns, while valid, do not fully appreciate the changes that AI will introduce. Liability under the law is fundamentally concerned with assigning responsibility for wrongdoing.[1] If someone is hurt through the actions of another, it is only reasonable that the second person be held responsible and that he or she help or pay the injured person. The assumption, of course, is that a person has done something and that a person will be responsible. The driver. The babysitter. The doctor. AI introduces another possibility. There is an accident, but all the people involved have acted responsibly and have not caused it. A machine, acting independently of people, caused it: Lt. Commander Data steers the Enterprise into an asteroid, Johnny Five from *Short Circuit* crashes a truck in Montana, etc.

What if Google Car hits someone? We assume that the operator of a car is ultimately responsible for any wrongdoing the car causes; this makes sense, as today's cars only move due to the direct actions (or inactions) of their drivers. As chapter 3 discusses more thoroughly, that will not be the case when Google Car and other autonomous cars appear on the market. So if Google Car hits someone while in autonomous mode, there is likely no human operator who has committed a wrong. And if the car were functioning properly, then Google is unlikely to be liable. Google Car may—or may not—be liable. The danger of applying our current liability laws to AI is that because they assume a human actor, they may not be broad enough to assign liability to a robot actor, potentially leaving accident victims with little to no recourse to receive compensation for their injuries or losses.

This chapter explores how our liability laws will need to change to address liability that either cannot be assigned to a human or must be distributed somehow between AI and people. I first provide a quick review of the forms of liability that will be most relevant to AI before explaining how those forms of liability will affect and be affected by AI. Given the liability that doctors expose themselves to every day, it makes sense to illustrate this by using the last portion of the chapter to look at AI surgeons, briefly explaining their development and then discussing liability issues that have already appeared and that may appear in the near future.

Theories of Liability

For over 200 years, the United States has addressed liability concerns related to new technologies with a complex hierarchical system of federal, state, and local entities. The system has changed enormously to accommodate new technologies, including mass production of goods, mechanized transportation, and electronic communications systems.[2] Many of the theories in place that address liability for disputes involving existing technology will also be relevant in addressing liability among AI.

Fans of *Robopocalypse* please take note: intentional torts—like trying to kill every person on the planet—are unlikely to become a serious issue with AI, at least not for a long time.[3] Battery, assault, false imprisonment, and other torts that require intent will not impact AI. The AI set to become available to consumers in the foreseeable future will be incapable of intent. The programming will permit AI to "intentionally" make decisions and act, but those will be limited to its established functions: driving from one location to another, performing surveillance for police, performing medical operations, etc. Although these examples put AI in a position to hurt people—like an autonomous car driving into a bystander—such actions will be incidental to its decisions. The AI will not attempt to hurt people, nor will it act with a substantial certainty that it will hurt people, standards of intent that are typically required under law.[4] If AI is programmed to intentionally inflict harm on people, the programmers would be responsible as they intended to cause that harm. The robot was merely the tool. Having said that, AI that assists police forces or the armed services (which are discussed in chapters 6 and 7, respectively) and that may be required to inflict harm is unlikely to be held to the standards of intentional torts in much the same way that police officers and soldiers are not.

Rather, AI is most likely to be involved with causes of action (claims or legal charges in court) that are accidental in nature: negligence, contractual violation, defective product liability, etc. For convenience, I'll summarize a few theories of liability that will be most relevant to AI:

Negligence

In general, negligence follows the reasonable prudent person standard. If a reasonable prudent person would avoid doing something (like throwing watermelons into the highway) because he or she believes that doing that might result in harm, then doing that "something" is negligent.[5] In other words, if you're doing something that a reasonable prudent person in the same or similar circumstance would avoid, you're acting negligently. This contrasts with intentional torts like assault or battery in that negligence does not require intent. You don't have to try to cause harm to be liable for negligence. The idea is that every person owes a certain level of care and consideration to the well-being and property of others no matter the situation: walking on the street, preparing food, throwing a baseball, etc. If someone does not maintain that level of care and as a result another person is harmed, the first person is negligent and liable.

Contractual Violation

Under the law of contracts, when two parties enter into a contract and one party does not perform his or her obligations under that contract, the other party is entitled to money or some other form of recovery. The actual form and amount may be stipulated in the contract itself, or if not, a court may determine the form and amount based on what the injured party has lost. A contract does not even have to specifically state one party's obligations, so long as the other party could reasonably rely on the first party to perform that obligation.[6] For example, if a wedding party reaches an agreement with a limo company to provide transportation from the church to the reception, although it may not be stated in the contract, the wedding party reasonably relies on the limo company to have insurance for its driver and vehicle. If there's an accident and the wedding party is injured, under the contract, the limo company will be treated as if it had insurance.

Defective Product Liability

Under defective product liability law, parties who manufacture or distribute harm-causing products are liable for the injuries or damages caused by those products, typically to the extent that it is feasible to prevent such injuries or damages[7]—that is, to the extent that preventing injuries and damages was both technologically and commercially possible.[8] For example, if adding a shoulder harness to a car's seat belt system costs less than the costs of accidents prevented by the shoulder belt, then the shoulder belt should be included in the car.[9] There are three basic defective products claims: manufacturing defect, defective design, and insufficient warning.[10] However, when injury is caused by a manufacturing defect, meaning the manufacturer did not actually produce the product in accordance with the manufacturer's own specifications, feasibility is not an issue and the manufacturer will be found liable.[11] Car accidents caused by defective brakes are among the most well-known examples of this type of "strict liability." Under defective product liability producers, distributors, and anyone else in the commercial distribution of a product may be found liable for harm caused by that product.

Malpractice

Malpractice is essentially a higher standard of care for professionals in certain fields. When a professional violates that standard of care, it creates liability for that person. Lawsuits involving malpractice are essentially negligence cases with special rules.[12] For example, a court has described medical malpractice cases as "a kind of tort action in which the traditional negligence elements are refined to reflect the professional setting of a physician-patient relationship."[13] Because negligence and malpractice are closely related, I do not address malpractice below, but it will certainly come up in liability claims involving AI.

Vicarious Liability

Vicarious liability, which is sometimes called *respondeat superior*, states that employers will be held liable for the actions of employees if those actions are performed in the scope of their employment.[14] I question how frequently this doctrine will be used because negligent supervision of the AI is a more accurate description of the likely scenarios where a business owner is relying on AI to perform functions that employees would otherwise carry out. However, if an employer reasonably relied on the AI's ability to perform a task, it might be easier for a plaintiff to win a court case by arguing that the AI should be treated like a human employee than by arguing that the employer negligently supervised the AI. Although this chapter does not discuss vicarious liability, chapter 4 briefly considers it in the context of a store using an AI babysitter to watch shoppers' children.

When Is AI Liable?

AI is going to force us to expand our idea of what can be liable. If a person is not liable, a machine may be. But in what scenarios will AI be held liable instead of a human being?

Liability Involving AI

As use of AI becomes more widespread and assigning liability due to accidents or incidents involving AI becomes more commonplace, there are essentially four scenarios that will occur:

1. *The person using or overseeing the AI is liable.* When an accident involving AI is caused by the person who uses or oversees the AI, then liability is properly assigned to that person. There are numerous possible situations in which this could occur:
   - An AI surgeon is operating on a patient, installing a new hip, when the machine suffers a mechanical error in mid-surgery. The human doctor overseeing the operation does not properly monitor the procedure and cannot step in to complete the hip replacement, causing the surgery to conclude early and forcing the patient to have a second hip replacement procedure.
   - AI construction workers demolish the incorrect wall, causing a building to collapse, because the human supervising the AI gave improper directions, indicating that the AI machines were supposed to demolish that wall.
   - The person overseeing AI passenger planes fails to notice that duplicate landing instructions were given to two planes, causing a midair collision.

   Any lawsuits filed against a human being meeting this description will most likely claim negligence (or malpractice, if that is applicable in the profession) as the reason why that person is liable. This category of liability involving AI acknowledges that there will continue to be situations in which human beings are required to oversee the

AI. However, there will probably be a significant schism in terms of liability assigned to human beings between AI that is adopted primarily to convenience people and AI that is adopted primarily to improve human performance. That is, people who use AI that primarily improves human performance will be required to monitor the AI and will be held liable if the AI causes an accident. People who use AI that primarily makes life easier for human beings will not be expected to monitor it and will face lesser—if any—liability for accidents caused by the AI.

Of course, the AI in question can perform both functions. For example, 90 percent of car accidents are caused by human error,[15] meaning that autonomous cars have the potential to greatly reduce the number of car crashes. However, as most drivers believe that they are good drivers,[16] the safety of autonomous cars is unlikely to sway consumers. Rather, car purchasers are more likely to buy an autonomous car because they like the convenience, are among a population for whom driving is a burden (visually impaired, physically impaired, elderly, etc.), or are young and don't believe driving is that big a deal anyway (as it decreases the available time for texting with friends).[17] So although autonomous cars may decrease accidents, injuries and deaths, for liability purposes it would be a mistake to believe that the primary purpose of autonomous cars is to reduce injuries and death. Most people will use an autonomous car to read, sleep, text, play Halo, or do things to use (or waste) the time they would otherwise spend driving; what they will not do is monitor the AI carefully. Once autonomous cars are common, it will seem somewhat ridiculous to find people liable for failing to monitor their autonomous cars because people won't monitor their autonomous cars. (This idea—assigning liability to "drivers" of autonomous cars—is discussed more thoroughly in the next chapter, as there are currently several state laws that would do just that.) In contrast, AI surgeons are not intended to increase human free time or make life more convenient. Rather, they are intended to improve upon human surgical results,[18] as they can make incisions, enter bodies, and flex at angles in ways that are impossible for human beings.[19] Human doctors that do not closely monitor their AI surgeons should be found liable for malpractice. The AI will not be not purchased to create more free time for them.

For similar reasons, courts looking to assign liability will distinguish between automated and autonomous (i.e., AI) machines. Automated machines carry out fixed functions, like an elevator or robot on a factory assembly line, but that is all; they do not adjust their functions (or not very much) to respond to new information without further programming. Even though a human being does not stand next to automated machines monitoring them, there are ultimately human beings, the programmer or maintenance personnel, responsible for their actions because the machines are incapable of making decisions themselves.[20] If an automated machine causes an accident—and the machine's design or production is not faulty—there is likely a human being that should have found a problem before the accident. However, autonomous machines are capable of making decisions, meaning that a human being may not have been responsible at all.

2. *Another person interacting with the AI is liable.* We have actually already seen this. In 2010, Google engineers reported that Google Car had been rear-ended by a human-driven car.[21] Liability claims that arise from the actions of a person interacting with the AI will be mostly identical to current liability claims that arise from the actions of a person interacting with any other machine or person. The biggest differences are likely to be in the predisposition of insurance adjusters or juries, which may (or may not) be more sympathetic to a person rather than a smart machine.

3. *The manufacturer, distributor, or another party in the chain of distribution is liable.* When an accident involving AI is caused because the AI malfunctions, is incorrectly programmed, is poorly constructed, is damaged during distribution, or because of any reason related to something going wrong with the product itself, liability is properly assigned to a party somewhere in the chain of distribution: designer, manufacturer, store where bought, etc. Basically, claims of this type will either be a standard defective product liability claim (the AI product did not work as intended) or a claim of negligence (the party preparing or delivering the AI did not exercise due care, and the AI was damaged before the consumer acquired it). A few examples will be useful:

   - An AI chef malfunctions and freezes while cooking a steak, burning the rice and ultimately the house where the robot was cooking.
   - The portion of code that controls an AI store clerk's ability to identify items that require an ID (cigarettes, alcohol) is accidentally deleted in one robot, leading to teenagers purchasing a case of Kiwi-Lemon Mad Dog 20/20, getting drunk, and causing an accident.

- The instruction manual for an AI snowplow fails to state that temperatures below −15 degrees can cause the robot to stall, and numerous accidents occur when these machines stall in the middle of northern New England roads during a freak, but not unprecedented, May blizzard.
- During delivery, a flying police drone is damaged, causing its engine to stall mid-flight and crash into a house near the police station.

The first scenario could either be a manufacturing defect or a defective design, and further investigation during a trial could identify that more clearly. The second scenario is a clear example of manufacturing defect: one particular robot deviated from the manufacturer's specifications for its programming requirements. The third situation illustrates insufficient instructions or warning. In contrast, the final example demonstrates negligence on the part of a distributor, which distinguishes this from a traditional defective product liability claim, but still removes the ultimate human user from responsibility for the damage. The human consumer would also have a contractual violation claim against either the manufacturer (for failing to use a safe delivery) or the distributor (for failing to safely deliver the AI), in violation of the purchasing agreement, which assumed the product would be safely delivered.

Although there is plenty of room for developers and manufacturers to deceptively advertise their AI products, I have chosen not to address it in this section because there will be little change in that area of consumer protection law. For example, although a developer may design a robot to provide child care, that company may be reluctant to advertise it as a nanny. If a child is hurt while the robot is nannying, the victim's family could claim the company falsely advertised that the AI provides effective child care. To avoid that charge, the developer might market the AI as a toy, even though the AI was designed as a child-care robot and parents are likely to use it for that eventually.[22] Claims of liability in this scenario would rely on what is "reasonably foreseeable" to create liability for the manufacturers. If manufacturers or developers can reasonably foresee that their products will be used for a certain purpose, the product's design must reflect that use, regardless of how it's advertised. So if a company tries to advertise that a squirt gun is not intended for squirting other people, the company is not relieved of its duty to make sure the squirt gun can squirt other people safely. Kids are going to squirt each other with a squirt gun no matter what the advertisements say. AI will not force changes to legal assumptions in those cases.

The first three types of liability involving AI that we have discussed do not deviate from standard liability claims that already exist under current laws. However, the fourth type is possible only through technology that makes decisions independently of human beings.

4. *The AI is liable for its actions.* In a few scenarios, AI will be treated as a human for liability purposes because it will have legal duties. AI will retain liability for its actions when the following circumstances occur: (1) the AI causes damage or injury; and (2) all of the responsibility for the damage or injury cannot be attributed to some combination of a human being or a party in chain of distribution. At times, when these circumstances occur, the AI will have performed *exactly* like it was supposed to, but will have still caused damage and should be liable. Here are a few practical examples to explain:

   • Two autonomous cars are driving on a highway, the first in the left-hand lane slightly ahead of the second car in the right-hand lane. They are traveling at reasonable speeds. The people inside them are not closely monitoring the vehicles but are wearing seat belts. A deer jumps into the middle of the left-hand lane. In less than a second, the AI in the first car makes the safest decision available and veers into the right-hand lane to prevent a collision with the deer. The AI in the second autonomous car has less than a second to act, but it chooses the safest decision available, veering to the right, which avoids the first car but forces a crash into a tree there. The people in the first car are fine. One of the passengers in the second car has a broken arm, but no one disputes that more people would have been injured if the second autonomous car had tried to stop in its lane (because it would have hit the first car, injuring the people in that car too) or swerved to the left (because it would have hit the deer and injured more people in the second car). The autonomous technology in both cars operated correctly. The cars were manufactured properly and did not suffer any relevant damage prior to purchase or prior to the accident.

   • An AI surgeon is performing a routine gallbladder removal, when a previously undetected weakened artery ruptures. The AI surgeon begins to repair the artery. The human doctor who is monitoring the surgery concurs with the AI's decision and plan. The AI surgeon is able to mend the artery, but only after the patient has suffered significant blood loss. Due to the internal damage, the AI surgeon decides not to remove the gallbladder at that time. The human doctor agrees. The patient must spend an extra two weeks in the hospital to recover from the artery repair before the rescheduled gallbladder removal. During an investigation after the surgery, it is determined that although the patient was thoroughly examined prior to the surgery, the weakened status of the vein was undetectable prior to the surgery. The investigation also confirmed that the human doctor monitoring the AI surgeon provided sufficient oversight and was adequately prepared to terminate the AI and step in to perform the artery repair if that proved necessary. The human doctor correctly noted, however, that the AI surgeon was taking the proper actions to repair the artery and save the patient. A diagnostic of the AI technology indicated that it operated correctly during the surgery. The AI was manufactured properly and did not suffer any relevant damage prior to the surgery.

- While working on a demolition site, a construction AI removes a beam. According to the blueprints of the building—that the human foreman had access to—that beam was not a load-bearing beam. Unfortunately, the beam in question was, in fact, carrying much of the weight of the remainder of the standing building. The rest of the building collapses, causing serious injury to people on the street adjacent to the demolition site. Subsequent inquiry reveals that the beam likely became load-bearing due to a combination of the ground shifting during the life of the building and the initial stage of the demolition. However, the inquiry also clearly demonstrates that it was impossible to know prior to the accident that the beam had become load-bearing and that the human foreman had provided appropriate oversight of the construction AI. A diagnostic of the AI reveals that it operated correctly during the demolition. The AI was manufactured properly and did not suffer any relevant damage prior to the demolition.

As these examples demonstrate, there will be numerous instances when there is an accident and the person monitoring the AI is not at fault, other people interacting with the AI did not cause any damage, the AI was designed and manufactured and delivered properly, and the AI was not damaged prior to the incident. Only the AI is responsible. Our laws do not account for that.[23] So whereas we have legal models now that can address the first three scenarios, we have to come up with new models or ideas to address liability that rests with AI.

A quick word about finding AI negligent: It is likely injured plaintiffs will allege negligence against AI. It is a natural claim to make (particularly for crafty lawyers) when human decision making is replaced by AI decision making. If humans must act reasonably and prudently, why shouldn't AI? And it is possible that these claims will gain traction. But I believe that such claims will probably be more appropriate against the programmers of the AI, as they programmed the AI to make unreasonable, imprudent decisions.[24] But even if the AI itself is found negligent, we are still left with the question of what happens when the AI is responsible, which requires a new model.

What Should Happen When AI Is Liable?

Once our laws properly assign liability to AI, we still need to determine who pays when the AI is responsible for damages. This will become a public policy decision, as legislators and judges will have to weigh interests of economic development (encouraging research and development of AI, encouraging the sale of AI) against fairness (victims should be properly protected and compensated for damages and injuries that are the fault of AI). There are essentially three options:

1. *The owner pays.* This is a simple rule, and one that is similar to liability associated with owning a dog or having kids. If you own AI, you should assume the risk that comes with it. And "assumption of risk" is a common defense to many allegations of liability in court, meaning essentially that the plaintiff accepted any possible liability voluntarily and released the other party, the defendant, from that liability.[25]

   I expect that this theory will be particularly popular among the manufacturers of AI early in the development and sale of AI, when customers bring lawsuits involving damages caused by early AI. The argument will essentially be that a particular type of AI is new and not completely defect free; although the company believes it is generally a safe product, the consumer purchased it knowing that the AI was new and that there are some risks inherent in that. Although some courts may find this persuasive at first, I suspect that it will be a losing argument for the most part. In the same way that our legal system currently expects a certain level of functionality and safety from all products on the market, the legal system will expect that a company's willingness to sell AI products indicates that it is guaranteeing that product satisfies a similar level of safety. The consumer is not actually assuming a great deal of risk by purchasing and using that AI.[26]

I admit that assigning liability for AI to the owners is a *plausible* public policy decision—it is simple, easy to enforce, and appears to have a certain amount of fairness, at first glance. Indeed, as chapter 3 explains, preliminary state laws governing autonomous cars essentially assign liability to car owners, or someone standing in the owners' place, as the "operator." But I do not believe that assigning liability to owners is the *best* public policy decision. First, it does not reflect the reality of what often occurs when AI causes damage or injury; the owner frequently will have had nothing to do with it. Unlike a dog or child, there is no expectation that the owner of AI be responsible for the robot at all times. That is one of the benefits of the technology; it can make responsible decisions for itself, unlike say a dog or (many) children.

Second, if the owner is liable, ownership of AI is discouraged. If part of the purpose of public policy is to encourage the development and commercial success of new technologies like AI, forcing potential consumers to assume all the risk is counterproductive. Even if consumers purchase insurance to protect themselves from that risk, there will still be the threat that insurance will not cover all the costs of liability in an accident. Owners in that scenario will have to pay for the damages of their AI. This places a disincentive on owning AI and, indirectly, a disincentive on developing AI, as there will be fewer consumers for it.

2. *The developer or manufacturer pays.* Like assigning liability to the owner, assigning liability to the developer or manufacturer is simple, easy to administer, and seems fair, at least at first glance. It even appears similar to strict liability: the manufacturer and designer did not intend for that accident to happen, so therefore, the AI has deviated from the manufacturer's specifications. However, it suffers from the same flaws as liability assigned to owners.

In the examples described above, the AI behaved as it was supposed to and frequently made the best of bad scenarios. Faced with an inevitable accident, the autonomous car chose the best possible accident; faced with a surprise mid-surgery injury, the AI surgeon addressed the artery damage and permitted the patient to survive.

Forcing the manufacturer or developer to pay damages for a properly working machine does not reflect the fact that they did not do anything wrong. It would be like requiring the manufacturers of wrecking-ball companies to pay for the buildings they demolish. The product is supposed to demolish those buildings. The AI is supposed to make the decisions discussed above.

Additionally, and most seriously, potential developers and designers will be discouraged from inventing new forms of AI if they believe their inventions will force them to pay damages even when their product works correctly. Potential manufacturers will feel the same way about producing AI that creates liability for their perfectly made machines. This policy might protect the relatively small percentage of people that are injured by well-functioning AI, but will damage everyone else due to lost development and AI advances.

3. *The AI pays.* As strange as it sounds to force the AI to pay, this is a legitimate option, and the one that I think is best. I don't mean to say that we should start paying AI a working wage—even C-3PO didn't have a take-home salary or savings account. But there are ways to provide a reserve of funds that pays settlement or restitution costs that are owed by AI, including requiring a certain level of insurance on the AI—which state legislatures are already requiring for autonomous cars—or adding a liability surcharge to any purchase of AI in order to create a government- or industry-maintained reserve that becomes available when AI is found liable. The regulation of these methods of paying for AI liability is discussed more in the next chapter.

AI Insurance versus AI Reserve Fund

It's important to understand the difference between AI insurance and an AI reserve fund. With a system of mandatory insurance, every individual AI-enabled machine that requires insurance would have a pool of potential money in case of liability-related payments. Assuming the AI was engaged and responsible, any settlement payments would be limited to the insurance proceeds; the owners and manufacturers would never pay more. With the reserve system, an entire class of AI would have a single pool of money to use in case of liability-related payments. All settlement payments would come from the pool, not from owners or manufacturers directly. Both AI insurance and an AI reserve fund are funded by a combination of owners, manufacturers, and developers paying in one form or another, either by splitting payment of the insurance policy or by splitting the surcharge cost at the time of sale.

The point of either system would be to give victims a way to recover financially for their losses while not actually making owners, users, developers, or manufacturers liable for the total potential losses associated with AI. By making the AI pay, through insurance or a reserve fund, the liability more accurately reflects what happened, as it was the AI that made a mistake. We also would not actively discourage consumers from buying AI or manufacturers from making AI. Because the liability and responsibility to pay rest solely with the AI, the humans involved—the owners, developers, and manufacturers—know the cost beforehand. They will not need to budget for unexpected lawsuits. The insurance or the reserve fund protects them from that.

However, what might be an issue for plaintiffs and victims is that AI will not always satisfy traditional liability claims like negligence, contractual violation, defective product liability, etc. In the scenarios I outlined, traditional liability claims would not apply. When the AI exercises appropriate care and performs as it was intended, negligence and defective product liability are losing arguments in court. Similarly, contractual language can be written so that the party offering the AI's services does not promise to provide certain results, but rather will provide properly functioning AI. That contractual language would make allegations of contractual violations unsuccessful in a court case involving AI in the scenarios above.

Instead of relying on traditional liability claims, it might be smarter to operate either AI insurance or the AI reserve funds similar to how worker's compensation insurance works. Worker's compensation requires a lower level of proof, so it is easier for workers to get paid, but the payout is lower than a potential court judgment.[27] Having an established insurance policy or reserve fund for AI-caused liability would remove the need to assert a traditional claim of liability. Rather, a plaintiff would only have to show (1) actual injury or loss, and (2) reasonable proof that the AI did it. In return for this lower standard of proof, plaintiffs would have to accept lower financial payment, in order to maintain the solvency of the insurance companies and reserves. As with worker's compensation insurance, this would help both those who suffer injury or damage from AI and those who own or create AI.

Liability of Robot Doctors

As doctors, surgeons, and other medical practitioners are among the most frequent subjects of liability claims, a review of how AI will affect liability associated with surgery will be useful to look more closely at how we'll sue AI. We actually are not as far away from widespread AI surgery as most people believe. Some people who have reviewed surgical practices believe that surgery today is more like "automatic surgery." Most of the work is done by residents and the "master" surgeon is frequently not in the room.[28] An AI surgeon, therefore, would not frustrate many veteran surgeons by limiting their hands-on involvement. It is true that human-controlled robotic surgeons have been much more popular in the last 20 years, and cyber surgeons (robotic systems that permit human doctors to operate on patients far away) have generated more excitement recently than AI surgeons.[29] However, autonomous surgeons have been developed and used in surgery, and I believe they represent the natural evolution of robotic surgical systems.[30]

Brief History of Robots and AI in Surgery

AI surgeons have made great strides in the last 25 years, progressing from essentially automated tools to self-directed surgeon-assistants. Robotics first appeared in surgery in 1985, when Doctor Y. S. Kwoh and his colleagues used a Puma 560 industrial robot to perform neurosurgical biopsies with greater precision.[31] The group of doctors modified the Puma robot to hold a fixture next to the patient's head so drills and biopsy needles could be inserted at a desired location for neurosurgery.[32] In other words, the doctors used the robot to hold the right spot in a patient's head while the surgeon drilled into the skull and removed a brain sample. Later in the 1980s, surgeons were able to use the Puma in this way to perform accurate resection (i.e., removal) of deep-seated brain tumors, which were previously inoperable.[33] The Puma 560 proved to be a catalyst for robots in surgery. In 1988, Doctor Brian Davies led a team of surgeons that used the Puma 560 to perform a transurethral resection of the prostate, removing prostate tissue to reduce its size.[34]

Puma (which stands for Programmable Universal Machine for Assembly) is a line of industrial robots designed for "medium-to-lightweight assembly, welding, materials handling, packaging and inspection applications."[35] It was originally produced by Unimation, Inc., the first industrial robot manufacturer,[36] with support from General Motors in the 1970s.[37] During the 1980s numerous automobile plants adopted it, and Pumas are still in use in many research labs today.[38] All Pumas, including the Puma 560, are essentially robotic arms that can swivel and flex in multiple directions, the perfect stereotype of a manufacturing robot seen in every news story about robots in factories.

Westinghouse Electric Corporation (which later became CBS Broadcasting) purchased Unimation, Inc. in the 1980s and eventually decided that robots were not designed for use adjacent to people, withdrawing its permission to doctors to experiment with the Puma 560.[39] This led to the development of Probot by Dr. Davies and his colleagues at London's Imperial College in 1991.[40] Probot was designed specifically to perform transurethral resections of the prostate, much as Dr. Davies had used the Puma 560.[41] Unlike the Puma 560, however, the Probot operated with little direction from the surgeon, relying instead on preprogrammed movements.[42] However, Probot's autonomous functions and independence in surgery were a hard sell to other doctors, and the Probot was not widely used. "Doctors just didn't feel comfortable with the idea," says Justin Vale, a consultant neurological surgeon at Imperial College and a fellow at the Royal College of Surgeons.[43]

While Davies developed Probot in London, Dr. Howard Paul (a veterinarian) and Dr. William Bargar (an orthopaedic surgeon), Dr. Peter Kazanzides (a robotics engineer), and Brent Mittelstadt (a biomedical engineer) founded Integrated Surgical Systems, Inc. in Sacramento, California, and created Robodoc, a robotic system that can core out the femur for a hip replacement with greater precision than a human surgeon.[44] Their goal was to carve out the bone cavity so that it would more precisely match the shape of the artificial hip replacement. Prior to Robodoc, surgeons used a mallet and hammer to chisel out a hole in the bone, and then cement was used to affix the replacement and compensate for any irregularities between the hole and the replacement joint. Robodoc uses sensors to detect the size and shape of the femur, after which, the doctor, using a computer terminal, enters the type of implant and the correct position within the femur.[45] Using this information, Robodoc digs the cavity in the bone autonomously, with little input from the surgeon, whose only role is to hold the emergency-off button.[46] Robodoc became the first commercially available surgical robotic system and is still in use today.[47] As of 2010, it had performed more than 24,000 procedures worldwide, and newer models have expanded to knee replacements as well.[48]

Probot and Robodoc represent the school of thought in medicine that robots can and should perform more surgical procedures with human oversight, but not necessarily with direct human control. This school of thought is leading the way toward greater reliance on AI in surgery. Nonetheless, Davies admits that this idea makes some doctors uncomfortable: "Although surgeons had thought that this autonomous feature was desirable, their unease at being just observers of a procedure that was largely in the control of the robot programmer soon became apparent."[49]

Dr. Davies believes that a class of robots known as surgical navigation systems addresses some of the concerns expressed by doctors who want human doctors to remain in complete control of surgeries.[50] Beginning in 1991, he began work with his colleagues at Imperial College on a robotic assistant, Acrobot (for active-constraint robot), that (in fact) actively constrains surgeons to cut accurately within a safe region.[51] Acrobot features a "hands-on" robot design, in which a force-controlled handle is placed near the end of the robot arm, which is manipulated by the human doctor during surgery. The robotic arm is what acts to actively constrain, essentially providing oversight, to the human performing the procedure.[52] "These robots are designed to hold the surgeon's hand in the operating theatre, not take over the operation. [They] can be an enormous help, preventing surgeons from making mistakes," said Justin Cobb, a professor at Imperial College.[53]

Despite these efforts at introducing autonomous surgeon robots and restraining robots in the operating room, the more dominant voices in the medical field have attempted to retain the human surgeon's full control. There are some good reasons for that. Even advocates of autonomous surgery robots concede that, at least for now, autonomous robots are less suited to soft-tissue surgery—that is, surgery on muscle, fat, skin, or other supporting tissues—as the tissue can change shape as it is pushed or cut.[54]

Rather, robot-assisted surgery and procedures are much more widely used and developed than autonomous surgeon robots. For example, robotic surgical systems act as a robotic extension of human doctors in the operating room. Da Vinci is perhaps the best-known example today, both because it was the first widely available robotic system and because it is used today in a variety of different procedures, including prostatectomies, cardiac valve repair, hysterectomies, cholecystectomy, and oral cavity resections.[55] Da Vinci is a type of robotic surgical system known as telemanipulator, a comprehensive "master-slave" surgical robot (meaning the robot "slave" is directly controlled by a human-controlled "master" computer) with multiple arms remotely controlled from a console with video and computer enhancement technology.[56] Its benefits include increased dexterity, improved visualization of the surgery (i.e., the doctor can get a better view of the area or areas where surgery occurs while remaining in his or her operating position), physical stability of the surgery, and the ability to perform the procedure remotely. Da Vinci, and other similar telemanipulators, makes surgeries that were extremely difficult, or even physically impossible, a reality.[57]

Da Vinci traces its origins to a collaboration between Scott Fisher, PhD, with NASA, and Dr. Joseph Rosen, a plastic surgeon at Stanford University.[58] In the mid- to late 1980s, they envisioned "telepresence surgery," which would permit "telecommuting" surgery by integrating interactive virtual reality with surgical robots. This team presented their research to Phil Green, PhD, and his team at the Stanford Research Institute, which used this concept to develop a telemanipulator for enhancing nerve and vascular anastomoses in hand surgery. With funding from the U.S. Army, engineers and doctors at SRI International (the same group that did the initial research that eventually produced Siri) led by Dr. Phil Green and Dr. Richard M. Satava developed this telemanipulator into the Green Telepresence System.[59]

The U.S. army had become interested in this technology because it would permit a field unit to place a wounded soldier in a vehicle with surgical equipment that a nearby M.A.S.H. unit could operate. [60] The Army provided funding for several different lines of research in this area, one of which led to the development of a table-mounted robotic arm controlled by the operating surgeon to manipulate an endoscopic camera, dubbed Aesop (for Automated Endoscopic System for Optimal Positioning), by Computer Motion, Inc. Aesop eliminated the need for a camera-holding assistant and became the first surgical robotic device approved by the FDA in 1993.[61] In 1997, Intuitive Surgical, Inc., licensed the Green Telepresence System, extensively reworked it, and rereleased it as da Vinci.[62] In 2003, Intuitive Surgical bought out Computer Motion and decided to release only da Vinci going forward.[63]

The advances made in telemanipulators by the U.S. government and the private research it supported have made possible the modern state of laparoscopic, or minimally invasive, surgery. Since the first minimally invasive surgery was performed in 1987 to remove a gallbladder, surgeons and engineers, working together to improve robotic surgical systems, have vastly expanded the procedures available, largely due to advances in robot-assisted surgery.[64] Because of these advances, the procedures lead to shorter hospital stays, faster recoveries, decreased pain, and better postoperation immune system reactions.[65]

And make no mistake: many of these benefits are only possible due to the robots. For example, because minimally invasive surgery, by its definition, seeks to *minimally invade* the body, the surgeon is not able to view the entirety of the area where he or she is operating. Telemanipulator robotic systems like da Vinci use multiple cameras to create three-dimensional images for the doctor to view while operating, even though there are only minimal incisions in the patient's body.[66] Additionally, physiological tremors—low tremors found in most people—transmit through surgical instruments, making delicate incisions difficult if not impossible.[67] As explained above, this is a specific issue addressed by tele-manipulators. And although da Vinci is the most widely used—having performed over 200,000 procedures in 2012 alone[68]—there are numerous other models, including Sensei X[69] and Epoch,[70] that have appeared commercially or in labs and feature the same basic telemanipulator model.

Similarly, there are other robots—either in development or currently available—that do not fit neatly into the telemanipulator classification but still follow the principle that human doctors should always be in control of a procedure. These include CyberKnife Robotic Radiosurgery System and Raven. CyberKnife is a frameless robotic system that removes tumors and lesions from the brain using focused ionizing radiation.[71] Raven is perhaps the most interesting recent development in the field of robotics-assisted surgery, and even in AI surgery. Raven, a robotic surgical assistant designed at the University of Washington, is intended to act quasi-independently of the human surgeons they assist. According to Gregory Hager, a computer science professor at Johns Hopkins who specializes in robotics: "The opportunity is to go from what humans can do, to doing things that are really superhuman ... And to do superhuman surgery will require robots to have enough intelligence to recognize what the surgeon is doing and to offer appropriate assistance."[72] Assuming Raven or another robot like it succeeds in performing those auxiliary functions independently, that could be the technological advance that pushes widespread adoption of AI in the operating room.

With that development, then, the three major types of robots that have been explored in surgery have led or will lead to AI surgeons, in one fashion or another:

1. *Autonomous Surgeons:* Probot and Robodoc are actual autonomous robots performing surgery. The benefits of autonomous surgeons include high-precision incisions and maneuvers, fast reaction speeds mid-surgery, and the ability to work on long surgeries without getting tired.[73] Although at this point they are limited to certain types of procedures, as their technology and programming are not able to compensate for variables introduced by soft tissue, they represent surgery without human doctors and make human doctors uncomfortable.

2. *Surgeries with Robot Oversight:* Acrobot represents the idea that human beings can perform the surgery but benefit from having a

robot oversee and limit it. Although this does not lead to AI surgeons in and of themselves, Dr. Davies seems to admit that Acrobot represents a compromise technology until doctors are more willing to accept autonomous robots in surgery.

3. *Surgeries performed by robots controlled by human doctors:* Da Vinci is the most popular and well known-robot in this category. However, as Raven demonstrates, the eventual goal many researchers have identified is to create "smart" robotic surgical assistants that can act in concert with, but independently of, human doctors.

Peter Kazanzides, whose Integrated Surgical Systems developed Robodoc, believes that "surgeons would not be afraid of autonomous technology if they felt it was beneficial for their patients and if they felt they understood it and could control it."[74] As doctors become accustomed to using robots in surgery, the trend suggested by Raven is not surprising: in the long run, doctors will start to rely on AI while operating despite their past reluctance because they will see that "superhuman" medical care is possible with AI technology.

AI surgeons also address a concern identified by the medicinal robotics industry: patients are not able to make an informed decision about surgeons, as there is not a satisfactory system to compare goods ones with bad ones. With a robot operating, patients know what they're going to get.[75]

Cases Involving Robotic "Doctors"

Despite the fact that robotic surgery systems and AI surgeons have existed in one form or another for 20 years, there are few court cases involving robotic surgery systems or AI surgeons. However, these cases hint at the complicated issues that will become apparent as AI surgeons become more common in hospitals.

The most infamous of these cases is *Mracek v. Bryn Mawr Hospital*, 363 F. App'x. 925, 926 (3d Cir. 2010), which involved a patient, Roland C. Mracek, who underwent a prostatectomy at Pennsylvania's Bryn Mawr Hospital in June 2005 after receiving a diagnosis of prostate cancer.[76] His surgeon intended to use the da Vinci to perform the surgery, but during the surgery, the robot displayed "error" messages and the human surgical team was unable to make the robot functional. The team completed the surgery manually with laparoscopic equipment.[77] One week later, Mracek found a large amount of blood in his urine and was hospitalized. Thereafter, he continued to suffer erectile dysfunction, which he did not have prior to the surgery, and has suffered severe groin pain.[78]

Following these injuries, Mracek sued Bryn Mawr Hospital and Intuitive Surgical, Inc. for monetary damages, citing multiple liability claims involving strict product liability and negligence. The hospital was voluntarily dismissed. Intuitive made a motion for summary judgment, which the trial court granted, noting that Mracek failed to establish causation between the injury he suffered and the robot's alleged malfunction.[79] The circuit court affirmed the trial court, noting in part that Mracek did not demonstrate that da Vinci was defective, that the defect caused his injury, or that the defect existed at the time the da Vinci left Intuitive's control.[80]

This case points to a couple of issues that will be relevant as AI surgeons become more common. First, *Mracek* stands out because it features a robotic surgical system that did not function and was not able to complete the surgery. Fortunately, the human surgical team was able to finish the procedure without the da Vinci. Unfortunately, Mr. Mracek suffered injuries, which he believed were both caused by the surgery and avoidable had the da Vinci been able to finish the prostatectomy.[81] Although the court ultimately found that there was no evidence of a connection between the malfunction of the da Vinci and Mr. Mracek's evidence, it is also important to note that the prostatectomy was successful.[82]

This speaks to the need for qualified and prepared surgeons to monitor AI surgeons. AI surgeons will have malfunctions and errors in the same way that today's robotic surgical systems have those problems. Although AI surgeons will hopefully be more reliable, if the AI fails, human surgeons need to be ready to step in and either stabilize the patient or complete the surgery, as the surgical team did for Mr. Mracek.

The second issue from *Mracek* that impacts AI surgeons deals with expert witnesses. The trial court noted that because da Vinci is so complex, a plaintiff would need an expert witness to help the jury understand whether or not the robot had a defect.[83] But *finding* expert witnesses can be extremely difficult for plaintiffs alleging liability in these circumstances; the court would not permit a doctor who had performed many robotic surgical procedures to provide expert testimony about the prostatectomy because the plaintiff could not show that the doctor had sufficient technical knowledge.[84] This suggests that the plaintiff would have needed someone who has specialized knowledge about the da Vinci *as well as* specialized medical knowledge in the field, permitting only three types of witnesses: a medical expert from the manufacturer, a medical expert from a competitor with specific knowledge of how the technology should operate, or a doctor who uses the robotic surgical system and also has sufficient training and experience necessary to make him or her knowledgeable about the robot.[85]

Although there are other robotic surgical systems, none of them exists as a true competitor to da Vinci, essentially giving Intuitive a monopoly. That eliminates the first two potential expert witnesses that could explain to a jury how the robotic system injured the plaintiff, as no Intuitive engineer or doctor is likely to testify as an expert witness against the company, and there are no competitors to provide alternative engineers and doctors.[86] Although expert users are a possible option, the court in *Mracek* barred an alleged expert user.[87] How does a plaintiff recover against the manufacturer of a robotic surgical system for injuries caused by the robot when an expert witness is needed to explain what happened to the jury, but there are no expert witnesses for the plaintiff to rely on?

Admittedly, other courts have been more lenient toward expert witnesses in trials regarding robotic surgical systems than the federal court in *Mracek*. For example, in *Gagliano v. Kaouk*, 2012-Ohio-1047, the patient suffered infection and other injuries following the "robotic removal" of his prostate. The jury delivered a verdict in favor of the defendant doctor, partly because of the expert witness used by each litigant. The court permitted the defendant doctor to rely on an expert witness who is certified to perform robotic prostatectomies; the court didn't require a higher standard. Of course, the court also permitted the plaintiff to use an expert witness who admitted he is not certified to perform robotic prostatectomies and did not know when robotic prostatectomies became standard.[88] So perhaps the leniency in this case will not be used as a model by other judges.

Regardless, the importance of expert witnesses is likely to increase when hospitals begin to use AI surgeons, and the issues highlighted by *Mracek* will continue to be issues to one degree or another. Although there is no reason to think that there will be only one manufacturer of AI surgeons, there is also no reason to think there will be more than one. We just don't know. If one company comes to dominate the market, a plaintiff trying to prove that the AI was defective will run into the same lack of expert witnesses that Mracek encountered. Even if more than one manufacturer is viable within the market, there will likely be few engineers and doctors between those manufacturers that will be well versed enough in the *other* companies' AI that they would qualify as expert witnesses. It's also possible that because human doctors will only oversee the AI surgeons rather than operate with them, a doctor with ample experience managing AI surgeons will not qualify as an expert witness. This would make it almost impossible for a plaintiff to recover against the manufacturer of an AI surgeon if the AI did something improperly during the surgery, never mind if the AI did *nothing* wrong in the surgery but is still liable.

Expert witness testimony was also the key consideration in *Williams v. Desperito*, C.A. N09C-10-164-CLS (Superior Court of Delaware, 2011). In that case, Dr. Thomas J. Desperito performed a laparoscopic prostatectomy using the da Vinci on the plaintiff, Robert C. Williams Jr. Following the surgery, Mr. Williams learned that his femoral and/or obturator nerve was injured during the prostatectomy, causing leg weakness and related injuries. Based on this information, Mr. Williams sued Dr. Desperito for medical malpractice. Among the evidence Mr. Williams hoped to present was the testimony of Dr. Myron Murdock, who would have addressed the ethics of advertisements used by St. Francis Hospital (where the surgery occurred) to promote its da Vinci system. However, the court prevented Dr. Murdock's testimony, stating that Dr. Murdock was not qualified to present expert testimony on ethical questions facing hospitals as they advertise robotic surgical systems. The court also stated that Dr. Murdock's proposed testimony would not even be helpful to the jury because Dr. Murdock proposed to say both that Dr. Desperito had not performed enough surgeries with the da Vinci to be competent *and* that St. Francis could advertise Dr. Desperito's services on the da Vinci despite his limited experience.[89]

Basically, the hospital benefitted from Dr. Murdock's confusing and contradictory testimony. He wanted to say, "The doctor was woefully unprepared, but the hospital was entitled to advertise his inexperience as expertise." The court was prepared to permit the first part of that testimony, but not if it included the second part. Hospitals using AI in the future should expect medical malpractice suits, but should not expect all expert testimony to be disqualified for such strange doublespeak.

Revisions to Laws Governing Liability Associated with AI Surgeons

Additionally, hospitals and other health-care providers should have policies in place governing the use of AI, physician requirements, and patient expectations. Doctors and their institutional employers can use these policies when making decisions about incorporating AI into their procedures and practices to help ensure that the AI complements and improves their services. If there are court cases alleging medical malpractice related to the use of AI, judges and juries will rely on these policies to help determine if medical malpractice actually occurred.

AI surgeons themselves may be the target of medical malpractice claims too. The medical malpractice claim exists to acknowledge that the medical profession relies on the judgment and independent analysis of the doctors and other workers in that field and that the traditional negligence standard does not adequately ensure those things.[90] But if the human beings in the profession begin to rely on the judgment and independent analysis of the AI surgeon, it will only be a matter of time before an injured plaintiff and his lawyer claim that the AI should have exercised greater judgment and analysis. This results in the issues and questions— Who is liable for AI mistakes? Who pays?—discussed above. The laws regarding negligence and medical malpractice will need to adjust to incorporate AI into medicine.

Revising Liability to Protect Accident Victims as Well as AI Developers

For developers and manufacturers to properly address risk management, they need an accurate idea of what their risk is. For developers and manufacturers of AI, that's impossible right now. The liability claims that are available to would-be plaintiffs—most importantly, negligence, contractual violation, and defective product liability—do not accurately reflect AI's capability to make decisions absent human direction. Judges, juries, lawyers, and insurance companies cannot predict with any reliable accuracy what will trigger liability for the creators and makers of AI.

The existing models for liability may not even be appropriate for AI, as AI can do everything right and still produce a negative outcome. The same is true of human beings making decisions, and that is one of the reasons why we have insurance for ourselves as drivers, as lawyers, as doctors, etc. A similar model may be necessary for AI, although insurance companies may be reluctant to insure a product, with its potential for manufacturer and design errors, as it would a human being, particularly when that product complicates such coverage by making human decisions.

The complexity of lawsuits against AI is also daunting for injured plaintiffs. Determining exactly who is to blame could prove to be time consuming and technically difficult. Was the injury caused by the actions of the human overseeing the AI? By the AI's defective design? By a manufacturing error? By damage caused to the AI? By the AI itself during its normal operations? Plaintiffs will need expert witnesses, but finding those may be difficult.

Providing an easier path for victims to recover monetarily for AI-related injuries that defines and limits the liability for developers and producers of

AI addresses these concerns. Depending on the form, that could be done through government action. However, the developing regulations governing autonomous cars, which is arguably the most advanced AI nearing the market, do not address this liability, instead assigning liability to the "operator," as the next chapter explores.

Part II

Must a Robot Obey Orders from a Human Being?

CHAPTER 3

The Uniform Artificial Intelligence Act and the Regulation of AI

While lawsuits involving AI will commence whether or not there are laws directly governing AI, state legislatures are already beginning to draft and pass legislation addressing the most prominent early example of AI in the marketplace: autonomous vehicles. Although not quite as slick as Knight Rider and KITT, Google Cars, the fleet of test Priuses Google has equipped with autonomous technology, have widely publicized this AI in the last few years. The proposed and enacted statutes, which Google Car has prompted, prove that inasmuch as nature abhors a vacuum, lawmakers might hate vacuums more. If there is an unlegislated topical issue, legislators will fill that empty space.

With that in mind and foreseeing numerous AI products entering the marketplace in the next decade, this chapter looks at how states are governing autonomous cars and what that suggests about how states will regulate other forms of AI in the future. Before getting to that, though, it is useful to discuss how another period of seismic new technologies was regulated and legislated: the Industrial Revolution. In many ways, this provides a potent illustration of how a new technology can affect average people both before and after regulation. It is useful to remember the Industrial Revolution and the laws that came out of it, which we now take for granted.

Please bear in mind that "regulation" is a broad term, encompassing almost any law, administrative rule, or ordinance that provides governmental requirements or controls.[1] The municipal zoning laws discussed in chapter 5 are also regulation. This chapter focuses solely on actual, proposed, possible, and likely requirements that governments may place on the consumption of AI. Who may use it? What are their obligations? What capabilities must manufacturers provide in each product? What must manufacturers tell consumers? In other words, this chapter focuses on the regulations that users and manufacturers will first encounter with AI.

Regulating New Technologies

The parallels between the Industrial Revolution and forthcoming AI developments underscore the need to effectively legislate AI. A quick examination of the laws the Industrial Revolution inspired and the reasons to regulate AI will be helpful.

What Laws Came Out of the Industrial Revolution?

The Industrial Revolution, lasting from approximately 1790 to 1860 in the United States, is the quintessential period of technological change that disrupted lives and the economy. The Industrial Revolution introduced the factory system, bringing workers together under supervision to produce goods using a central source of power; this contrasted with previous forms of industry that had no central power.[2] In the early 19th century, most manufacturing in America was done in individual homes or, in larger towns, in a local mill or shop. Visitors to many towns would find a wide assortment of small manufacturers: lumber mills, grist mills, weavers, shoemakers, tanners, tailors, and others. But these varied forms of manufacturing were somewhat deceiving. The actual output of these establishments was quite small and only served the immediate area in and around the town.[3] This was how manufacturing had been done for hundreds of years.

Compared to those centuries of stagnant industrial development, the Industrial Revolution featured technological change at warp speed. This led to farms that needed fewer workers to produce more food and factories that required a greater number of workers to manufacture more goods.[4] In 1810, Henry Clay told Congress that a "judicious American farmer, in his household way manufactures whatever is requisite for his family."[5] But even by then, this was already on the fast track to American myth. Samuel Slater had already opened the first industrial mill in the country, greatly increasing the speed at which cotton thread could be spun into yarn. Although manufacturing technology had advanced little in the previous 800 years,[6] within the next 50 years, manufacturing and the lives of average workers would change tremendously.

In practice, what that meant was that factories required a constant supply of workers but had little incentive to police their workers' ages, safety, or working conditions. In contrast, factory owners had every reason to keep wages low, as well as the ability to do so. There was no precedent for the factories that appeared during the Industrial Revolution. They employed hundreds and thousands of people, producing far more goods than home manufacturing and tradesmen could. But employees had no laws to ensure that factory owners did not take advantage of them; consumers had no laws to ensure that factory owners did not pass hidden costs, like health or environmental damage, on to them.[7] This led to widespread use of child labor in factories, where the youngest ages were as low as four.[8] Working hours in factories exceeded 10 hours per day. Aided by their attorneys and sympathetic judges and legislators, employers killed any potential workmen's compensation or unemployment insurance laws into the 20th century.[9] At the same time, state courts consistently ruled that an employer would be blameless for an employee's death or injury if another employee was partially responsible.[10]

Although the factories provided many new products at prices that would have been impossibly low before the Industrial Revolution, there was no way for us, as a society, to ensure that the benefits of the new technologies were legitimately widespread. Although cotton sheets were cheaper, were the workers paid fairly? Did the production ruin otherwise drinkable water? Did the fumes and other byproducts cause illness and death among those who lived near the factory?

Eventually, there were laws that attempted to adequately distribute the benefits created by the technological advances of the Industrial Revolution. They ensured the creation of a large industrialized middle class in the 20th century but were slow in coming, partly because of lobbyists and lawyers employed by factory owners and industrialists throughout the 19th century. But by 1879, states had begun passing child labor laws, and a national prohibition against the practice was enacted with the passage of the federal Fair Labor Standards Act of 1938.[11] The Fair Labor Standards Act also established a national minimum wage, $.25 an hour, for the first time.[12] Recognizing that many states were reluctant or unable to pass laws that effectively protected unions, the federal government began protecting employees' right to organize after the National Labor Relations Act became law in 1935.[13] Similarly, state efforts to limit pollution of air, land, and water resources in the 1960s led to the creation of the federal Environmental Protection Agency in 1970.[14] While these legislative and regulatory efforts were ongoing in the 20th century, employees also benefited from the enactment of federal regulations governing workplace safety.[15] These laws ensured that while the creators and owners of the factories, machines, technology, and patents of the Industrial Revolution retained their well-earned fortunes, the workers powering the industrial economy also retained some of the financial and material benefits.

Why Regulate AI?

AI poses the same problems created by the Industrial Revolution and other successful new technologies. AI products will create economic benefits for the economy as a whole, but the initial designers and manufacturers of these products will keep a substantial portion of that benefit. In the same way that manufacturers were able to employ whole families in their factories at the beginning of the Industrial Revolution with little requirements in terms of working conditions, hours, child labor, environmental externalities, or pay, AI products could enter the market without appropriate testing, safety mechanisms, or warnings. As well-funded 19th-century manufacturers were able to afford expensive lawyers to protect their interests in court, well-funded 21st-century AI manufacturers will be able to afford expensive lawyers to protect their interests, creating a need for regulations that establish rules for liability, user obligations, and manufacturer obligations.

So if the expansive American industrial middle class of the mid-20th century was only made possible by the child labor laws, minimum wage laws, legal protections for unions, and other legal reforms that more equitably spread the enormous economic efficiencies and material comforts introduced by the Industrial Revolution, new regulations are necessary to make sure the enormous benefits of AI will be equitably spread throughout the economy. Here are just a few scenarios where AI creates an obvious benefit while also possibly inadvertently creating a huge liability or economic loss:

- AI construction workers may reduce the cost of new homes, commercial buildings, and renovations by reducing or eliminating the salaries associated with those projects. But if there are no minimum standards for the AI, the owners of those properties have no reassurance that the AI workers will be supervised properly or are even qualified to complete the construction. Construction projects might cost less, but the ultimate occupants (who are not always the people purchasing the construction) may experience more problems with the building(s) or pay high repair costs.
- AI doctors may lower the cost of surgeries by reducing the number of high-paid medical staff. But if *no* qualified medical staff is

required to monitor a procedure, more patients will die or suffer serious mid-surgery injuries due to the AI.

- AI pharmacists may permit human pharmacists to interact with patients more, improving the standard of care and patient satisfaction. But if regular maintenance is not required, the consumers—who may have no idea that AI filled their prescriptions—may receive incorrect medications.

- AI vehicles may create more usable time for people with long work commutes, as they will be able to work, text, sleep, or read during their commutes instead of driving. But if liability associated with the vehicles is not regulated properly, courts may inappropriately assign liability to human "drivers" merely because they turned on the car, not because they caused any actual damage.

(Those scenarios do not include the intellectual property issues associated with awarding the ownership of a best-selling novel's copyright to a programmer when that person didn't write the novel, only the code of the AI that wrote the novel, a scenario discussed in chapter 8. They also do not include the likely job losses associated with AI, which is discussed in chapter 9.)

Ideally, regulations will address these situations and create requirements that minimize AI's potential harms while maximizing AI's potential benefits.[16] What regulations should not do, however, is protect the jobs of those displaced by autonomous technology. Some of the professions discussed in this book dismiss the effect AI could have on their fields because "government regulations" protect them.[17] Although there may be some truth to that, that is not the purpose of regulation, particularly with regard to new technologies that dramatically change business models, like AI. One of AI's primary benefits is that less labor will be required for all sorts of industries: construction, manufacturing, physical and medical care, etc. Job losses are expected and are not necessarily bad, assuming we can create a public system that addresses displaced professions.

So to summarize this chapter thus far, regulation of AI is coming, and that is good. And not only some regulation. Quite possibly a lot of regulation. As a more diverse array of AI products appears on the market, there will be more pressure on state legislatures to pass laws addressing those types of AI. We have already seen this with autonomous cars. The auto industry is still years away from a commercially available autonomous car, but legislation exists today to regulate it. This pattern won't necessarily repeat itself for every type of AI. Siri and Google Voice are unlikely to be regulated. Legislators may question the public safety imperative behind regulating AI writers, composers, or other AI that does not physi-cally interact with people and the world around us. But there will be enough AI in different fields that seriously affect human safety that the old Vulcan proverb "Infinite diversity in infinite combinations" will accurately describe the expected laws regulating AI. From that diversity, it is possible that an overarching law or laws governing AI in general will emerge, a Uniform Artificial Intelligence Act, which this chapter considers.

Because legislatures are already proposing laws regulating autonomous cars, they represent an ideal lens to use when viewing and considering the regulation of all AI. They indicate the issues legislators are concerned with and suggest how legislators will address them. There also seems to be a gen-eral consensus among decision makers—government administrators and legislators—that we must amend our automotive and highway laws because autonomous vehicles will change our assumption that only people make decisions. As David Strickland, Administrator of the National Highway Traffic Safety Administration (NHTSA), stated in late 2012: "Most of NHTSA's safety standards assume the need for a human driver to operate required safety equipment. A vehicle that drives itself challenges this basic assumption. This is also true of state efforts to govern motor vehicle safety. State highway safety programs overwhelmingly focus on preventing driver behaviors that are deemed unsafe such as speeding or impaired driving."[18]

Additionally, AI vehicles are approachable. Most people don't interact with surgeons, pharmacists, or lawyers on a daily basis. Almost everyone interacts with a car, bus, or vehicle every day, making the benefits (free time instead of driving) and harms (damages caused by autonomous vehicles) readily apparent. The consequences of regulating autonomous cars are therefore easy to appreciate.

State Regulation versus Federal Regulation

You'll note that despite the fact that federal laws and regulatory schemes would preempt any state action, I refer almost exclusively to state legisla-tion in this chapter, both existing and expected, although I discuss a pos-sible federal uniform AI act. There are several reasons for this.

First, states are much more aggressively seeking to regulate autono-mous cars than the federal government. This is not for want of trying by some people at the federal level. At the Autonomous Vehicle Seminar on October 23, 2012, David Strickland stated that the NHTSA would initiate research on emerging autonomous driving technologies before the end of the year.[19] Volvo has started lobbying for federal regulation.[20] However, despite the NHTSA's commitment to fund a multiyear $1.75 million research project to study real-world effects of autonomous technology in vehicles,[21] there is no passed or pending federal legislation.

Second, given the recent state of paralysis in Congress, it is unlikely that federal legislation addressing autonomous cars will become law anytime soon. Third, even if Congress passed a federal autonomous car statute—perhaps something akin to various federal consumer protection acts[22]—there will be state laws and regulations as well, as is the case in many industries where state and federal laws coexist. And in the same way that states regulate the automobile standards people are most familiar with—licensing, inspections, insurance, etc.—state regulations governing autonomous cars will be the regulations most people encounter.

Finally, because I believe that regulatory treatment of autonomous cars is useful in predicting future legislation of AI, it is difficult to predict how federal legislators will address AI when the only evidence we have regarding the regulation of autonomous cars is exclusively from state legislatures. Until Congress acts otherwise, the states are where the action is.

Driverless Drivers Licenses: Regulation of Automated Vehicles

Although I am cognizant of the problems introduced by regulating new forms of technology too soon, regulation of autonomous cars is not necessarily "too soon." Self-driving vehicles have been in active development for decades.

An Abbreviated History of Autonomous Vehicles

Although Leonardo da Vinci is sometimes credited as attempting to design the first automated mobile cart,[23] the idea of autonomous cars first gained popular attention at the 1939 World's Fair in New York.[24] General Motors created a ride called Futurama (from which the Matt Groening cartoon *Futurama* derived its name) that took passengers on a tour of the cities, roads, and infrastructure of America in the distant future—1960!![25] GM had created a diorama featuring miniature cars, buildings, and cities, each detailing aspects of life in 1960. In addition to serving as advertising for the interstate highway system that was still almost 20 years away, Futurama also told visitors that GM envisioned green parkways filled with self-driving cars.[26]

In many ways, this represented the public coming-out party for the "automated highway system" that automobile engineers, vehicle component manufacturers, government officials, individual inventors, and readers of speculative technology magazines had dreamed of during the

1930s.[27] GM's exhibit described a system of radio-controlled cars—featuring vacuum tubes, the most advanced technology at the time—that would maintain safe distances between cars, bumper to bumper, without direct human control.[28] Strangely, while advanced radio controls maintained safe distances between car bumpers, the Futurama exhibit relied on Luddite physical barriers between lanes to maintain safety on each side of driving cars.[29]

For a variety of reasons, not the least of which was World War II, GM did not invest a great deal of resources into its driverless cars or the automated highway system for more than 10 years. By the 1950s, wartime advances, such as radar, made the automated highway seem more plausible. In 1956, GM and Radio Corporation of America (RCA) were exploring a system of traffic control towers, installed periodically along major highways, that would electronically control cars.[30] In 1958, GM issued a press release stating it had achieved a driverless car: "An automatically guided automobile cruised along a one-mile check road at General Motors Technical Center today [February 14, 1958], steered by an electric cable beneath the concrete surface. It was the first demonstration of its kind with a full-size passenger car, indicating the possibility of a built-in guidance system for tomorrow's highways. ... The car rolled along the two-lane check road and negotiated the banked turn-around loops at either end without the driver's hands on the steering wheel."[31]

GM presented another concept for the automated highway system at the 1964 World's Fair, but, although public interest grew immediately following the fair, private investment in the research declined thereafter. As one engineer put it, "[T]here was sort of a hiatus as we came to limitations in terms of the technology that was available when measured against the economic feasibility."[32]

It should be noted that the automated highway system and autonomous vehicles are two distinct ideas, although they share a common goal: a car that does the driving for its human passengers. The automated highway system is "a technology that links highway and vehicles" by creating some type of actual interface—radio control, cables, etc.—between car and road.[33] Autonomous vehicles contain technology that can analyze their surroundings and direct themselves based on that input, regardless of the road underneath. Up until the 1960s, research had heavily favored the automated highway system, but beginning in that decade, institutional and governmental research shifted to autonomous vehicles as those entities began to develop unmanned ground vehicles (UGVs). Whereas GM and RCA researched transport vehicles, UGVs explicitly did not carry people.[34]

Through the 1960s and '70s, researchers at Stanford University made great strides in autonomous robot vehicles. By 1969, AI work funded by DARPA—the same defense department agency that spearheaded the development of the Internet and Siri—had led to the creation of the aptly named "Shakey," a wheeled platform with an ultrasonic range finder and touch sensors, which was connected via radio to a mainframe computer that controlled navigation and exploration tasks. From 1973 to 1981, Hans Moravec at the Stanford Artificial Intelligence Lab explored robot navigation and obstacle avoidance issues, ultimately creating a UGV named the Stanford Cart.[35] The Cart was a buggy equipped with a video camera, remote control, and a very long cable (apparently the AI Lab was located in the first Wi-Fi blind spot on Stanford's campus).[36] Eventually, it used the video camera to send images to an offboard mainframe that processed them and made decisions about the Cart's next move. It was both incredibly slow—the mainframe could take up to 15 minutes to make a one-meter move[37]—and an incredible achievement, becoming the first robot to successfully cross a chair-filled room without human intervention, albeit in five hours.[38]

At the same time, labs outside of the United States were making progress too. In 1977, S. Tsugawa and his colleagues at Japan's Tsukuba Mechanical Engineering Laboratory unveiled the first car that could process images of the road ahead, using two cameras and an analog computer. It was capable of speeds exceeding 18 miles per hour but relied on an elevated rail.[39] In the 1980s, German engineer Ernst Dickmanns, a professor at Bundeswehr University Munich, began making great strides in autonomous driving technology, eventually earning the unofficial title "Pioneer of the Autonomous Car"; his success is sometimes credited with permanently redirecting research from the automated highway system to autonomous vehicles.[40] In 1987, an autonomous vehicle of his design— equipped with 2 cameras, sensors, and eight 16-bit microprocessors— drove more than 50 miles per hour over a distance of 12 miles.[41]

During the 1980s, DARPA's Strategic Computing Program began to work extensively with the autonomous land vehicle (ALV). The ALV consisted of a jury-rigged all-terrain vehicle equipped with a color video camera and laser scanner, as well as data processing modules to process road data. In 1986, ALV's autonomous technology successfully drove more than 6 miles per hour over a course that was more than 2.5 miles long. In 1987, the ALV successfully navigated another course with varying pavement types, road widths, and shadows, measuring over 2.5 miles in length, while averaging 9 miles per hour.[42]

The progress of autonomous technology continued steadily from there, as private companies, governments, and educational institutes contributed to a growing field of successful autonomous vehicles. Using funding from the European Union, Dickmanns worked extensively with Daimler-Benz to experiment with autonomous technology, a partnership that produced an autonomous test model. Beginning in 1994, the test model could navigate normal freeway traffic in France and Germany.[43] In 1995, Dickmanns equipped a Mercedes S-class with autonomous technology, and it traveled nearly 1,000 miles from Munich to Odense, Denmark.[44] According to Dickmanns, about 95 percent of that journey was travelled autonomously, without human control.[45]

Also in 1995, Dean Pomerleau and Todd Jochem, engineers from Carnegie Mellon University, finished equipping a 1990 Pontiac Trans Sport, dubbed Navlab 5, with a portable computer, a windshield camera, a GPS receiver, and supporting equipment. The computer took pictures of the road ahead, used those pictures to orient itself, and then produced steering commands to propel Navlab 5 and maintain the vehicle's lane position.[46] Using this technology, Pomerleau and Jochem planned and publicized a cross-country trip, in which Navlab 5 would drive itself and the engineers from Pittsburgh to San Diego.[47] The trip proved to be a success, with Navlab 5 driving 98.2 percent of the trip (2,796.87 of the 2,849.13 miles) autonomously.[48]

DARPA's funding has sustained American research of autonomous technology for decades,[49] and that funding resulted in the DARPA Grand Challenge in 2004 and 2005. In 2003, DARPA announced a race scheduled for March 2004 that would require autonomous robots to navigate a 300-mile course of off- and on-road terrain between Los Angeles and Las Vegas. DARPA would award $1 million to the team whose robot could complete the course first in a defined time period.[50] Of the 106 teams that applied, 17 qualified to compete in the race on March 13, 2004, but not a single autonomous vehicle finished the course. The most successful entrant traveled only seven miles. What is remarkable about the DARPA Grand Challenge is the progress made between that first event in 2004 and the second event in 2005. Nearly 200 teams attempted to enter the 2005 race, the course of which crossed dry featureless lakebeds, railroad crossings, road intersections, and tunnels, all designed to confuse the sensors and processors of the vehicles. Of the 23 finalists, 5 teams finished the 140-mile course in less than 10 hours.[51]

The winning robot was designed by a team from the Stanford Artificial Intelligence Laboratory. Sebastian Thrun, the director of Stanford's AI Lab, is also a Google engineer. While helping to design the winning autonomous vehicle, Thrun began to experiment with autonomous technology through Google, creating Google's now famous fleet of Google Cars.[52] Since then, Google Cars have logged more than 400,000 autonomous miles on California roads.[53] Only one accident has been reported while the car drove itself, when a human-controlled car drove into a Google Car.[54] Thrun touts safety as one of the key benefits of autonomous cars. He believes that Google Car can eliminate as many as half of the deaths caused by road accidents today.[55] Other benefits include reducing congestion and maximizing fuel economy.[56] Industry experts now expect autonomous cars to be commercially available by 2020.[57] Assuming they have voice recognition software akin to Siri or Google Voice Search, we'll be able to talk with our cars like a favorite taxi driver.

State Regulation of Automated Cars

With autonomous cars poised to take to the highway and interact with human-driven cars en masse, in 2012 state legislators began to consider laws and regulations to address this technology in a way that was unnecessary when only sporadic defense and university research teams worked in this field. As of the end of 2012, only California, Florida, and Nevada had successfully legislated autonomous cars; the District of Columbia passed its Autonomous Vehicles Act of 2012 in 2013. However, several other states have considered or are in the process of considering legislation governing autonomous cars through the 2012–2013 legislative session: Arizona, Hawaii, Massachusetts, Michigan, Minnesota, New Hampshire, New Jersey, New York, Oklahoma, Oregon, South Carolina, Texas, Washington, and Wisconsin. Most of that legislation is similar to California's or Florida's, proposes studies of autonomous vehicles, or requests further input from the state's Department of Motor Vehicles or Transportation. Nevada is the only state that has created regulations to flesh out its legislation.[58]

However, some notable trends have begun to emerge, as the laws tend to:

- *Find that there is no legislation governing autonomous cars*— California and Florida, in passing laws addressing autonomous cars, specifically recognized that each state "does not prohibit or specifically regulate the testing or operation of autonomous technology in motor vehicles" or "the operation of autonomous vehicles."[59] In doing so, the states admit the major point this book seeks to impress: existing laws don't properly address AI. The acts passed to address autonomous vehicles will, by design and necessity, alter the bedrock legal principle that only people make decisions.

- *Create a liability shield for the original vehicle manufacturer if another party installs autonomous technology (e.g., Google)*—The legislature in Florida has also recognized that autonomous technology introduces AI to vehicles that original manufacturers should not be held liable for.[60] DC has a similar provision.[61] It is unreasonable for Toyota to be held accountable for errors caused by the AI designed and installed by Google. This is such a commonsensical premise that Nevada and California did not even address it in their legislation.[62]

- *Create a license endorsement to operate an autonomous vehicle*— The Nevada law[63] establishes a system in which the right to "operate" an autonomous car is attached to a person's driver's license, like the endorsements or license classes typically needed to operate a motorcycle or school bus. The failed and pending legislation in Arizona,[64] Hawaii,[65] New Jersey,[66] and Oklahoma[67] contemplate the same system. This system is in conflict with the admission made in California and Florida that the existing laws governing people-driven cars do not properly govern autonomous cars. The endorsement system assumes that the holder of the endorsement will operate the vehicle, when that is not the point of the autonomous car. Rather, tying use of autonomous cars to a license endorsement is a tacit acknowledgment by these legislatures that we, as a society, aren't ready for the new laws that AI demands. The entire history of human laws has assumed that people make decisions. These legislatures have attempted to split the robot baby in order to make the shift to robot-made decision more palatable: Yes, the car will make the driving decisions for you, but you must have the right driver's license to enjoy robot-made driving decisions.

- *Require a human "driver"*—The legislation passed in the District of Columbia specifically requires that a human being be in the driver's seat, "prepared to take control of the autonomous vehicle at any moment."[68] So far, this requirement is not so much a trend (no state has considered such a strict provision through the end of 2012), but it is a worrisome introduction to the effort to properly regulate AI in cars. Whereas California and Florida have acknowledged that existing laws are based on a premise that is no longer valid and Nevada has passed a law that attempts to bridge the gap between the old premise and the new one, DC's proposed legislation appears to actively discourage the use of autonomous technology. It is the technological equivalent of having your cake and not eating it: "Feel free to use Google Cars in Washington, DC, but make sure you're not enjoying the experience." Although this might appear mostly identical to the endorsement system used or proposed by other states, the endorsement requirements are not so specific. An autonomous car requires an "operator" with the proper endorsement, but the individual's seating in the car is unspecified; she could sit in the trunk (assuming New Hampshire's lax seat belt regulations are in place). Even if she is in the driver's seat, Nevada, Florida, and California do not require that she be "prepared to take control of the autonomous vehicle at any time." These laws recognize the benefit of AI in cars—the driving decisions do not have to be made by a person.

That idea surely makes many people at least a little uncomfortable. My concern is that uncomfortable legislators in other states, who are not ready to embrace this new era of laws (as California and Florida appear to have done), will not choose the middle-ground license endorsement system, but will opt for the unreasonably restrictive DC system. That would make the benefits of AI in cars largely irrelevant and ignore the fact that these driving robots can make decisions as people can. Requiring people to be engaged fully as decision makers is not necessary.

- *Identify the person who engages the autonomous technology as the "operator"*—Florida included the following text in the bill it passed: "[A] person shall be deemed to be the operator of an autonomous vehicle operating in autonomous mode when the person causes the vehicle's autonomous technology to engage, regardless of whether the person is physically present in the vehicle while the vehicle is operating in autonomous mode."[69]

  California's laws states: "An 'operator' of an autonomous vehicle is the person who is seated in the driver's seat, or if there is no person in the driver's seat, causes the autonomous technology to engage."[70]

  And Nevada's regulations include similar language, noting that: "[A] person shall be deemed the operator of an autonomous vehicle which is operated in autonomous mode when the person causes the autonomous vehicle to engage, regardless of whether the person is physically present in the vehicle when it is engaged,"[71] and "For the purpose of enforcing the traffic laws and other laws applicable to drivers and motor vehicles operated in this State, the operator of an autonomous vehicle that is operated in autonomous mode shall be deemed the driver of the autonomous vehicle regardless of whether the person is physically present in the autonomous vehicle while it is engaged"[72]

It is good that these laws anticipate the likelihood that people will initiate the autonomous technology in their cars and do not prohibit it. Future autonomous cars in Florida, California, and Nevada may drop off children at school alone or return home after bringing their owners to work, issues that are addressed in more detail elsewhere in this book. Consumers will be able to enjoy the full benefits of autonomous cars when legislatures accept the full benefits of machine-made decisions. Florida, Nevada, and California have begun to do that by accepting the scenario in which a car is operated but the "operator" is not in the car.

At the same time, although the laws in Florida, California, and Nevada contemplate autonomous cars driving without people, it is not clear that the legislators viewed this as a good development. Typically, liability follows the operator of a vehicle; if a vehicle caused an accident, the operator is liable. Identifying the person who turns on the car as the "operator," regardless of his or her control of the car, creates a scenario in which courts and the law will incorrectly assign liability for accidents caused by the autonomous car, as the previous chapter discussed in more detail. The second provision from the Nevada regulations appears to state that specifically.

If two automated cars crash into each other while their "operators" are many miles away, how is either operator liable, regardless of which vehicle caused the accident? Legislators and courts will need to more carefully assign liability based on who or what actually causes an accident. Liability based on the person-operator will need to become less important in order to reflect what actually occurs on highways.

- *Distinguish between autonomous vehicles used for testing and those operated by consumers*—California,[73] Nevada,[74] and Florida[75] included different requirements for vehicles used to test autonomous technology as compared to those autonomous vehicles sold to customers. Legislation in New Jersey, Arizona, and Oklahoma also proposed to differentiate between test vehicles and commercially available vehicles, although the exact details are not spelled out in the legislation. Limitations placed on test vehicles that are not placed on commercially available autonomous cars include restricted geographic areas and roads; the requirement that a driver be behind the wheel at all times, ready to assume control of the vehicle; and enhanced insurance premiums. This represents another example of an intermediary step toward full acceptance of vehicles that are capable of making decisions. It is unusual for legislation concerning un-autonomous cars to address the testing of those cars in this manner. However, until autonomous technology becomes more common on highways, such differentiations between unsafe test AI and safer tested AI in vehicles represent a practical approach and a better alternative to DC's requirement that *all* autonomous vehicles, whether tested or not, must have a human being in the driver seat, always ready to assume control of the vehicle.

- *Require a mechanism that permits a human being to assume control of the vehicle easily*—The laws enacted by California[76] and Florida[77]—as well as the regulations promulgated by Nevada[78]— all require autonomous vehicles to have an easily accessible means to engage and disengage the autonomous technology, that is, a way for a person to start driving the car. The legislation passed by the District of Columbia also contains a similar provision. Although this appears similar to the requirement that a human being be prepared to assume control of the car, it is, rather, a basic emergency safety feature. Parachutes work very well; if they didn't, no one would skydive. But every parachute has an emergency chute, "just in case." This requirement is the autonomous car's emergency chute. Just in case something happens to the AI, a human being can drive. It's important to recognize that this does not detract from the effectiveness of the AI.

- *Require a system that alerts a human being that the autonomous technology has failed*—All three states that have passed legislation governing autonomous cars, but not DC, have included a provision that requires a system that will alert people in the car if the autonomous technology fails.[79] This is another emergency chute provision, similar to the requirement that a human being be able to take control of the autonomous vehicle. As with that provision, it does not detract from the effectiveness of the AI.

  California has added an additional provision: If the autonomous technology fails and "the operator does not or is unable to take control of the autonomous vehicle, the autonomous vehicle shall be capable of coming to a complete stop."[80] This provision further addresses the likely scenario in which the operator is not present in the car and adds a necessary safety provision. I hope that other states—Florida and Nevada included—adopt this idea.

- *Require disclosure stating what information is collected by manufacturers of autonomous technology while cars are in operation*—

California[81] requires manufacturers of autonomous vehicles or autonomous technology to provide a written disclosure to all purchasers of autonomous cars that describes what information is collected by the AI. This speaks to two concerns. The first is that manufacturers of autonomous technology will add driving habits and destinations to Internet activity as sources of consumer information that will permit marketing companies (particularly Google) to know increasingly personal and formerly private aspects of our lives. The second is that the AI in the autonomous technology will use that information to make personal decisions for you: it's Friday after work, so the car automatically drives you to your favorite bar, regardless of where you actually wanted to go. The first concern is widely spread among many products and is not specific to AI products. The second concern is one that other AI products will have to address. Many AI products that make decisions for you will be harmless. Facebook and Google already use primitive AI to suggest links, stories, and videos to you. But others will give people pause. If your grocery shopping robot decides you need more fiber in your diet, but selects a product you are allergic to, there will be serious repercussions.

- *Require a state agency (the Department of Motor Vehicles or the Department of Transportation) to enact regulations that will fulfill the goals listed above*—Every state legislation[82] discussed substantively in this chapter has addressed autonomous vehicles, but has directed various state agencies to provide the details that will effectively govern autonomous vehicles. Thus far, only Nevada has successfully enacted these regulations. In reality, this is how most states (and the federal government) address complicated issues that require attention to detail. Rather than letting the legislature get lost in the weeds of important regulatory matters that govern people's everyday lives—highway safety, environmental oversight, etc.— legislators permit the agencies that will have to oversee the operation of the regulations (and answer phone calls from concerned or confused citizens) to create the regulations themselves, subject to certain requirements: new rules are publicized, the public can comment, etc. I point this out because it illustrates that the states that have attempted to regulate AI in cars have done so using the same tools they would use to regulate people-made decisions. There is no reason why we can't regulate AI-made decisions in the same way.

What Autonomous Car Regulations Suggest about Regulating Other AI

The trends in autonomous car regulation listed above suggest the trends that will become relevant as governments seek to regulate other forms of AI, not the least of which is the discomfort many people will feel as they attempt to sort through a world where robot-made decisions will have to be recognized as having the same force as person-made decisions. There is an inherent danger in trying to design legislation too soon for new technology.[83] The telecommunications industry is still struggling with zoning ordinances passed in the late 1990s that assumed every new antenna would create a new eyesore, meaning that even when a smaller antenna replaces a larger one, a wireless provider (Verizon Wireless, AT&T, etc.) frequently has to appear before a town board, rather than merely obtain a building permit. However, as stated in the first chapter, in order to protect a prosperous middle class, as a society we need to learn to quickly modify our laws to adapt to technological advances due to the increasing speed of technological change. So although it is hazardous to speculate about the regulation of technology that has not yet fully developed, it is worthwhile to think through future laws and regulation governing AI using the current and attempted regulation of autonomous cars as a model.

Late action is likely worse than early action. To use telecommunications as an example again, even though the industry has been inconvenienced by overly stringent zoning ordinances, wireless communications have still expanded exponentially since the 1990s. So although those ordinances have been annoyances to wireless providers, they have permitted towns to control their development and character in the face of rapid technological change. Below are reasonable predictions based on autonomous vehicle legislation and existing legislative practices.

1. *From the absence of legislation governing AI, legislatures will produce many specific-AI laws before the development of a Uniform Artificial Intelligence Act.*

I suspect that initially state legislatures will approach each AI development individually. Autonomous cars will soon be commercially available, so legislatures are developing laws to govern autonomous cars. When AI surgeons become more common, legislatures will develop laws to govern AI surgeons. When AI lawyers become common, legislatures and state bar associations will develop laws to govern them. And so on. Each time, they will have to admit what Florida and California had to admit about autonomous cars: there are no existing laws that address AI. But as companies develop more AI, that admission will be less accurate. States may decide that they do not want a hodgepodge of laws and regulations governing AI differently in cars versus doctors versus lawyers. Certainly, those companies creating AI will want predictability and uniformity as they develop new AI. There will likely be some sort of Uniform AI Act, similar to the many other uniform acts that govern sales and transactions,[84] trusts,[85] controlled substances,[86] etc. This will provide overarching standards for AI and permit more specific laws governing particular types of AI as necessary.

This Uniform Act may also take the form of federal legislation, adding a layer of federal requirements to each state's requirements. Another alternative is that a federal agency, possibly the Consumer Protection Agency, adopts as a model law a Uniform AI Act, as drafted by an industry group or American Bar Association advisory council. In adopting it, that agency then requires states to adopt the model law, or a law substantially similar to it, in order to receive full federal funding from the agency or in order to avoid further federal oversight. The latter is not an unheard-of practice. For example, recently the United States Department of Housing and Urban Development, pursuant to the Housing and Economic Recovery Act of 2008, required that states pass the model Secure and Fair Enforcement for Mortgage Licensing Act (the SAFE Act), or a state law substantially close to it. The SAFE Act requires states to have a system in place for licensing and registering mortgage loan originators. If HUD determines that a state's mortgage loan originator licensing system is insufficient—that is, it does not satisfy the standards established in the model SAFE Act—HUD is authorized to implement and administer a licensing system for the state. Were the federal government to take a similar approach with AI regulation, states would have to adopt a model Uniform AI Act or a federal agency would implement and administer regulations governing AI that conform to the Uniform AI Act.

2. *Different systems will emerge to create funds for victims of accidents involving AI where manufacturers, developers, and human beings are not at fault.*

The Uniform AI Act and the specific-AI acts leading up to it will almost certainly devote considerable attention to liability. In what situations will the manufacturer be liable? In what situations will a human being be liable? What distinguishes human fault from AI fault? The laws in California, Florida, and Nevada suggest that first attempts to assign liability will be somewhat clumsy. They will broadly paint human users as "operators" for liability purposes, even when those users have not actually controlled the AI or made decisions that led to an accident. As discussed in the previous chapter, in its most basic form, courts and statutes assign liability to the person or persons, company or companies, who cause damage. Except in rare "Act of God" instances, our current concept of liability assumes a human being made a decision that caused damage. The emergence of widespread AI will force us to reexamine that. Hopefully, the evolution of AI laws will quickly recognize that liability must expand beyond human beings. Assuming that occurs, states will likely play with a few different mechanism to do this, including the systems discussed in chapter 2:

- *Insurance on AI.* Florida,[87] Nevada,[88] and California[89] require some form of an instrument of insurance, a surety bond, large cash deposit, or proof of self-insurance for autonomous vehicles, although the statutes vary as to how much coverage is required and whether that coverage is required for both testing and operating vehicles. The states that have proposed legislation appear ready to move in this direction as well, as the acts typically contain language that calls for a state agency to "establish requirements for the insurance that is required to test or operate an autonomous vehicle."[90] This requirement essentially mandates the creation of a pool of money for every vehicle, available whenever that particular autonomous car is in an accident. If such insurance were in place for when the autonomous car itself is liable, that would remove the need to inappropriately assign liability to the manufacturer, the human "operator," or a third party. A similar requirement for AI surgeons, construction workers, and pilots makes sense. That being said, a similar requirement for smaller AI-devices, where liability may not be as prominent an issue, does not make sense.
- *AI reserve fund.* For AI products that are not likely to be in a position to cause major damage to lives and property (e.g., Siri, AI housekeepers, AI writers, etc.), it may make more sense to assign a small surcharge by statute at the point of sale in order to create a statewide or industry-wide liability fund for accidents where the AI performed exactly as it was supposed to but there was still an accident. For example, if your AI housekeeper is responsibly dusting your expensive handmade Oriental rug on a clothesline outside and a bird poops on the rug, you could approach the AI reserve fund

for some sort of payment. If this system is implemented, a minimum claim amount will likely be necessary to preserve the reserve fund for people who really need it and to save the administrative cost of hearing requests for payments related to insignificant damages, like the $20 rug from Target that the bird pooped on. For those lesser damages, legislatures should decide that is within the assumed risk of the AI product.

Liability should not be a major concern for every AI product. Most AI that does not physically interact with people will not require much, if any, specific legislation addressing its liability. But for AI that performs physical tasks, legislators should think carefully about likely accidents in which the manufacturer isn't liable because the AI performed as intended and the person who turned on the AI isn't liable because he or she wasn't required to monitor it. In those scenarios, accident victims are potentially left without recourse despite serious loss or injury. The new laws governing AI need to protect those victims as well as the manufacturers and owners who have done nothing wrong.

3. *At least initially, the use of many forms of AI will require a license endorsement.*

Requiring a special permit or license to operate certain AI products is likely. Such licensing is particularly attractive when the function that is being replaced also requires licensing, as with driving. We can expect that states will require driver's license endorsements before permitting residents to use autonomous cars. Similarly, we should expect to see similar endorsements for AI doctors, AI pilots, AI forklift operators, and any other profession or function where the person replaced by AI already has a license or permit to perform the same tasks.

As I mentioned above, this reflects a compromise between a world in which only people make decisions and the world we're entering where machines and computer programs make decisions too. And licensing endorsements is not a bad first step for AI legislation. For legislators who are nervous about the use of AI technology and the widespread replacement of so many human jobs—which is a good concern to have—licensing ensures a level of human oversight and employment. Much of the AI technology that initially becomes available on the market will likely benefit from more oversight.

I expect initial specific-AI legislation to rely heavily on license endorsements where the AI's human counterparts also require licensing. I wouldn't be surprised to see some form of that requirement in the initial Uniform AI Act. But my ultimate hope would be for the license endorsements to largely drop as technology improves, as we are culturally more comfortable with AI decisions, as the consensus builds that the AI is likely better suited to the task it performs than the majority of people it interacts with, and as legislators realize that the skills confirmed by a license are the ones needed if AI stops working properly. For example, anyone who is already licensed as a forklift operator should be qualified to turn on and monitor an AI forklift because if he needs to take over for malfunctioning AI, the skills he demonstrated in obtaining the license are the skills he will rely on to operate the forklift. It will eventually make as much sense to require special AI licensing from the people who turn on and monitor a forklift as it does to require special licensing from the people who drive a car with an automatic transmission.

Having said that, where specialized knowledge is required and human lives are particularly vulnerable, license endorsements will probably continue, meaning that some level of human oversight will continue as well. In other words, license endorsements will be required to use AI that is primarily intended to improve human performance, but not for AI that is primarily intended to make human lives easier. To use examples from last chapter, while the purpose of autonomous cars is both to improve highway safety and create more free time for the person who doesn't have to drive, the purpose of AI surgeons is only to improve surgery safety. The doctor's free time is largely irrelevant, and it is important to have a trained surgeon who is comfortable interacting with the surgical robot in case the AI malfunctions. For AI of this sort, license endorsements are not likely to disappear anytime soon.

4. *Initially, human oversight will frequently be required, but eventually only AI whose primary purpose is to improve human performance will require human oversight.*

Although requiring a license endorsement before drivers can use an autonomous car suggests more human oversight than is actually necessary, states have thus far mostly, and wisely avoided requiring *actual* human oversight of each autonomous car, the District of Columbia notwithstanding. Similarly, I expect there will be a few states that step beyond requiring license endorsements of relevant AI and require actual active human supervision of AI. In some cases, this is warranted; in others it will be unnecessary and will actively inhibit the growth of proper AI use.

As I've discussed, there are two groups of AI products: those whose primary purpose is to permit people to do other things and those whose primary purpose is to do people's work better. Autonomous vehicles are within the former group. Requiring people to actively monitor AI that is supposed to relieve them of the responsibility of monitoring —for example, autonomous cars, robot nannies, robot housekeepers, etc.—defeats the purpose of that AI. With regard to autonomous vehicles, Nevada, California, and Florida have chosen a prudent middle ground, license endorsements, that addresses the concerns of people who distrust machine-made decisions while also allowing the purchasers of this AI to enjoy its benefits. As explained above, I expect similar legislation initially about other AI whose sole purpose is to create more free time for its users, but I also expect that a small handful of more reactionary states will require "constant vigilance" in addition to license endorsements for this type of AI. However, slowly the fear of AI will dissipate, and endorsement requirements will disappear for these other types of AI.

On the other hand, products like AI surgeons do not exist to create free time for doctors. AI surgeons exist because they can meet the demands of surgery—particularly surgeries that last for hours in highly sensitive areas of the body—better than human surgeons. Humans become tired after the sixth hour of brain surgery; AI surgeons do not. A human hand may quiver at the wrong time during open heart surgery; an AI surgeon appendage will not. However, the requirements of the human doctor replaced by an AI cardiac surgeon are different than the requirements of the human driver replaced by an autonomous car.

If the autonomous technology in the car fails—triggering the mandatory alert system to notify a human being—a person in the car can immediately look at the road and determine an appropriate action, without much difficulty. Even if a human driver is not in the vehicle, the autonomous car can direct itself to the side of the road to stop. Active human oversight is not necessary. Let the "driver" read a book.

If the technology in an AI surgeon fails—triggering an alert system to notify a human being—the human doctor may be required to step in at nearly an infinite number of distinct points of the surgery. If that person has not monitored the robot's progress, is not aware of the robot's place in the procedure, cannot quickly identify the next steps of the surgery, or is unable to perform the remainder of the surgery to completion, the patient's life may be in danger. Even though he or she may not actively perform the surgery, the doctor must actively oversee the AI surgeon and the surgery. Specific-AI laws and the Uniform AI Act will need to reflect this.

As these laws are drafted, there will be discussion as to how we draw the line between AI that primarily eliminates human responsibilities and AI that primarily eliminates human error. What AI can operate freely and what AI requires our supervision? Take AI construction workers, for example. Permitting AI to perform most construction crew tasks—bulldozing, welding, crane operating, etc.—potentially eliminates the need for humans to perform those responsibilities. But is it necessary for humans to oversee the robots that physically perform those jobs? If the autonomous technology in an AI wrecking ball fails mid-demolition, is it necessary to have a human being directly monitoring it? It is possible that one or several foremen, monitoring the machines on a worksite but not attached to any particular machine for oversight, are able to successfully take control of that wrecking ball to complete the demolition project, or at least to ensure that no one gets hurt. Legislators will have to ask similar questions about other AI performing human tasks that are not necessarily as complex as surgery but that can cause great damage or loss of life if the technology fails.

5. *Although initial legislation will frequently identify the human operator, even when that label is inappropriate to the type of AI, eventual legislation will be more nuanced when identifying the operator for liability purposes.*

Legislators are going to be very concerned about the person who is the "operator" of AI for the same reason that they have been interested in identifying the "operator" of autonomous cars: traditionally, liability for machines attaches to the person responsible for that machine, that is, its operator. But the operator label will not be appropriate in many instances. The operator of an autonomous car, as Nevada, Florida, and California define it, is frequently going to be the operator of the car in the same sense that someone pushing the button for the 13th floor is the operator of an elevator. New laws should treat autonomous cars as the operator when they are driving for liability purposes because they are in fact driving; actual people are not. Under current state law, the human operator of an autonomous car turns the technology on but does not actually operate it. The label and ensuing liability are incompatible.

Nonetheless, initial specific-AI laws will almost certainly identify the operator so that courts, consumers, and businesses have that label in mind. It shouldn't be surprising if the first Uniform AI Act also identifies AI operators. But for much of the AI that primarily eliminates human responsibilities, the human operator label will disappear as inappropriate to the technology. However, laws are likely to always identify the person or persons who are the operators of AI that eliminates human error. Those operators will be responsible for monitoring that AI, if not actually using those robots. But our laws will need to be more nuanced. The person who turns on an AI surgeon or pilot is not necessarily the person who manages it.

6. *Legislation will consistently distinguish between AI for testing purposes and AI for consumers.*

In the same way that I accept laws that distinguish between autonomous cars for testing and those that are for consumers as a practical requirement, I believe it is good that we will see many different versions of this differentiation among laws governing other forms of AI. Autonomous cars are novel now, and dangerous. But the underlying concepts are not complicated: you tell the car where to go, and it drives you there. As time goes on, these cars will become commonplace, and the technology will be widely accepted. Because the underlying concept is so simple, we will eventually reach a point where the autonomous technology will be mundane. Future autonomous test cars that seek to improve that technology will not be treated any differently than the already tested and approved autonomous cars.

Legislation addressing AI that performs other physical functions—pilots, nannies, construction, etc.—will almost certainly distinguish between the AI that is well-tested and AI that is new. Some of these AI will also feature such simple premises that the technology used will be so widely accepted that testing future models will be less an issue of new functionality and more an issue of new features. The difference between the two is the difference between testing a plane with new engines and testing a plane with new passenger seats. The first may blow up and kill bystanders; the second may leave passengers with sore lumbar muscles. Once types of AI reach the stage of research, design, and product development when only new features are at issue, it will be unnecessary to distinguish between tested and untested models.

In contrast, other forms of AI may perform physical functions but will not have simple premises. AI surgeons will become increasingly complex, capable of performing all manner of repairs in the human body. AI drones will increasingly assume greater responsibility for patrolling streets and investigating crimes. These tasks are complicated. Much as we want bright human minds performing them, we will want well-designed and -tested machines performing them in the future. There will be many new designs, as designers upgrade earlier models to include greater functionality: early AI drones will monitor criminal activity for human police, but later models will make actual arrests. If early models are not thoroughly vetted, they may misanalyze data and fail to reduce crime. If later models are not thoroughly vetted, they may arrest inappropriately or violently. Legislatures will likely want to continue to differentiate between tested and untested AI of these types because the underlying technology will continue to improve to incorporate broader premises. It will be a long time, if ever, before one type of AI surgeon technology is almost universally used.

That is not to suggest that every type of AI will require test permits and operator permits, as appears to be the growing consensus on autonomous cars. The laws will consistently address tested and untested forms of AI that perform complex physical functions. The actual treatment will vary. For example, legislatures are unlikely to require manufacturers to obtain a test permit when developing AI police robots. But legislatures or municipalities might require certain levels of testing before police departments within their jurisdictions can use AI robots as part of police duties. Hospitals may encounter similar regulations governing AI surgeons. Boeing and AirBus, however, may have to obtain testing permits and licenses, like manufacturers of autonomous cars, before they can sell to airlines. It will depend on the product and, in some cases, current events surrounding the product. AI surgeons are in limited use now, and like autonomous cars there are no special laws prohibiting or regulating their use. That may change if cable news reports an accident or death.

7. Legislation will consistently require a mechanism that permits a human being to disengage the AI and easily assume control.

The requirement that every AI product come with a convenient off switch addresses two important issues. The first is the initial uncertainty and fear many people will feel about machines that can think. The second is the need to have an "emergency chute" in case there is an AI malfunction that necessitates a human operator. Although the first issue will gradually dissipate as use of AI products becomes widespread, the second will be a constant issue. Not likely one that will be on display in the news every day, but a legitimate concern, in the same way that emergency brakes are in every subway car. They aren't necessary every day, but you will want them when they are necessary. The need to have a human being take control of the AI in an emergency will make this requirement good public policy, while voters' initial uneasiness toward many types of AI will ensure that legislators include this requirement.

It goes without saying, though, that legislatures will largely restrict this requirement to AI that performs physical functions. AI that largely focuses on advising people, creative expression, or the production of other written materials will not require an emergency off switch. The need for a human being to take control of an autonomous plane is much greater than the need for a human being to take control of Siri or an autonomous music composition program. Siri would likely get peeved, anyway.

> 8. *Legislation will consistently require AI products to alert nearby human beings when the autonomous technology has failed.*

This is the natural complement to the requirement that AI products performing physical functions have an easy mechanism to let a human being assume control of the product. What's the point of permitting a person to assume control when the AI malfunctions if that person doesn't know when the AI malfunctions? The same reasons will undergird the legislating of this requirement: initial concern about thinking machines and the legitimate need to protect consumers from malfunctioning AI.

> 9. *Concerns about the collection of private and personal information will compel legislation that requires disclosure of the information AI collects as it operates.*

In contrast to the last two predictions, which were only relevant to AI interacting physically with people and the world, the prediction that legislation will require AI manufacturers to disclose the information collected about human users and beneficiaries is relevant to many types of AI.

Although legislators, policy makers, and consumers are rightly concerned about hackers breaking into personal e-mail accounts or laptops and stealing private information that way, the threat of corporately collected and lost private information is as big a worry if not more so. Consider the following timeline:

- June 6, 2005—Citigroup notified 3.9 million customers that UPS lost computer tapes containing their personal information, including Social Security numbers, names, account history, and loan information.[91]
- June 17, 2005—MasterCard officials announced that more than 40 million credit card accounts—belonging to MasterCard, Visa, and Discover cardholders—had been illegally accessed in 2004, when a hacker compromised a processing center operated by one of its vendors, CardSystems Solutions, Inc.[92]

- May 22, 2006—U.S. Department of Veterans Affairs announced that the personal electronic information of up to 26.5 military veterans was stolen from the residence of a Department employee, who had taken computer disks home without authorization.[93]
- March 28, 2007—The TJX Companies, Inc.—which operates several brand stores, including T.J. Maxx, Marshalls, and Bob's Stores—filed documents with the U.S. Securities and Exchange Commission admitting that unknown hackers stole 45.6 million credit and debit card numbers from its computers over an 18-month period beginning in July 2005.[94] Later reports indicated that the intruders accessed approximately 100 million customer accounts, which included customer names, purchases, credit and debit card numbers, and driver's license numbers.[95]
- August 1, 2008—The FBI arrested a former Countrywide Financial Corp. employee, alleging that over a two-year period he executed a scheme to steal and sell sensitive personal information, including Social Security numbers.[96] Further investigation indicated that the employee sold the information of as many as 17 million Countrywide customers. Eventually, others associated with the thefts were arrested, and courts issued prison sentences.[97]
- January 20, 2009—Heartland Payment Systems, a credit card processing company, announced that its data security system had been hacked.[98] Court documents later indicated that approximately 130 million credit card accounts had been accessed and compromised.[99]
- November 10, 2011—Due to defacement of its online forums, the Valve Corporation conducted an investigation that revealed hackers had accessed a company database with customer information, including names, passwords, video game purchases, e-mail addresses, billing addresses, and encrypted credit card information.[100] Later reports estimate that 35 million accounts were compromised.[101]

These events only represent the largest data security breaches since 2005, not all of them.[102] Nonetheless, these seven incidents compromised approximately 352 million accounts with personal information in about six years. And it was not as if there was a reliable pattern to the victims. They included older veterans, young gamers, bargain shoppers, credit card users, and homeowners with mortgages. Data security at both private companies and government agencies was breached, and it was not always on purpose. Even when data is protected from hackers, an errant UPS package can leak personal information.

The most sensitive personal data contained in the examples above include Social Security numbers and credit card account information, but records regarding purchases and other activities were also put at risk. This information reveals a lot about an individual, more than we realize. Target, for example, has developed an algorithm that uses about 25 products that can predict both *if* a female shopper is pregnant and *when* she is due to give birth. With this information, Target can send her coupons timed to specific stages of her pregnancy. Target's system is so good that recently an angry father confronted a Target manager because his teenage daughter had received advertisements for maternity-related merchandise. He thought the store was encouraging his daughter to get pregnant. The manager apologized and felt so bad about the incident that he called again to apologize to the father. However, the father confirmed that his daughter was, in fact, pregnant and that "there's been some activities in my house I haven't been completely aware of."[103] Andrew Pole, the architect of Target's marketing campaign targeting pregnant women, stated, "We are very conservative about compliance with all privacy laws. But even if you're following the law, you can do things where people get queasy."[104]

Which is why legislators will realize the importance of addressing information collected by AI manufacturers while people use their products: AI products will interact with people in many more aspects of their life, providing ample opportunity for companies to data mine and monitor your activities to collect information that can be used to sell future products to you. Stores like Target are more or less limited to monitoring your activities in their stores or on their web pages. Credit card companies are limited, more or less, to monitoring the purchases you make with their cards. Even Google, which arguably has the most sophisticated commercial system for monitoring hundreds of millions of individuals' search and online activities, is limited more or less to monitoring your activities on the Internet.

If you use Google to search for a local Target, drive there and buy a PlayStation with cash, and then buy gas on your way home with your MasterCard, those are three separate transactions that were observed by three different companies, with no way to directly connect all three. Google knows you searched for Target, but doesn't know if you went there or bought anything. Target knows you bought a PlayStation, but not how you found the store. MasterCard knows you bought gas, but nothing about the rest of your errand.

But if your Google Car drives you around town, Google knows you searched for Target, knows you drove to Target, and knows you stopped at a gas station on your way home. That information is available by connecting your Google account information to the AI in your Google Car. If Google is particularly interested in data mining (and not particularly interested in personal privacy), it could use the technology that permits you to give vocal addresses to Google Car to also pick out key words you speak in the car, like "PlayStation." That way, it has a better chance to connect all of the pieces of this errand and use them to sell other products to you in the future. And remember that Google Car and other autonomous vehicles will likely use advanced weak AI. We'll talk to our cars, give them personalities, and treat them like friends. We will be more likely to divulge information useful to advertisers in that frame of mind.

This possibility understandably frightens a lot of people, many of whom want to retain some semblance of a private life, parts of which are protected from inspection by companies whose products we happen to buy. I'm not inherently opposed to AI manufacturers compiling information based on how consumers use the AI, and I seriously doubt Google will attempt to record your phone conversations in your Google Car. However, our laws must reflect how we want information gathered by AI, how we want that information used, and what we want consumers told about that information. The California autonomous vehicle statute represents the idea that, at the very least, consumers should know what information their autonomous car collects. I think this will be the very minimum obligation that other legislatures impose on AI manufacturers.

Having said that, I take seriously the idea expressed in the introduction: we should fashion our laws so that the benefits of AI are widely spread among many people and are not limited to the few who design and produce the AI. I hope legislatures also consider proactively banning some data collection (e.g., monitoring conversations) that occur near AI that has speech-recognition software. And where data collection is not banned, it would be beneficial if a system is put in place where AI users receive a financial benefit from the data they produce while using the AI. Perhaps specific-AI laws can, as appropriate, require a licensing agreement at purchase. If the manufacturer will collect information while the product is used, it must pay the consumer for use of the data he or she creates.

10. *Legislatures will require that state agencies enact regulations that provide more specific requirements for AI and AI manufacturers.*

Despite introducing serious change to one of our fundamental legal assumptions, legislatures will address the governance of AI much as they already address the governance of everything else: regulatory code drafted and administered by state agencies. State departments of environmental protection draft and enforce regulations governing emissions, wetlands protection, and other aspects of environmental law. State departments of transportation draft and enforce regulations governing state highways. State departments of motor vehicles draft and enforce regulations governing cars, drivers, etc. These state agencies can also impose sanctions for breach of their rules.[105] Not surprisingly, it appears that legislators will task state departments of transportation and motor vehicles with preparing regulations that will govern autonomous cars. There is no reason to believe that other specific-AI laws will not continue this trend.

The larger question is whether states will create some sort of Department of AI or AI Board to administer the forthcoming Uniform AI Acts. State legislatures have enacted executive departments, councils, and boards to govern an enormous variety of regulatory areas, including midwifery,[106] guide dogs for the blind,[107] and purchasing from people with disabilities,[108] among many other topics. It would not be illogical or far-fetched for states to extend the same level of priority to AI, particularly if the technology becomes widely used. While a full-fledged regulatory department, akin to a Department of Environmental Protection, may be unlikely in some states, it is possible that every state will at least create a Council of AI under the state Council of Consumer Protection. This will solidify the centralization of AI governance that the Uniform AI Act represents.

These boards, agencies, councils, etc. would be able to promulgate specific-AI regulations that divide AI between those that perform physical functions and those that don't, as well as between those that eliminate human responsibilities and those that primarily eliminate human error. But they would also be able to promulgate general regulations governing all AI. This might give states a platform to quietly pass a law akin to "No AI may injure a human being or, through inaction, allow a human being to come to harm" without appearing on *The Daily Show* as crazy science fiction bureaucrats.

Autonomous Cars as the Future of AI Regulation?

The legislation governing autonomous cars provides an interesting, although small, sample of laws governing AI, providing a means to predict laws that will govern AI in the future. This chapter expresses how regulation of autonomous cars (and AI in general) falls short, how legislatures are attempting to address those shortcomings, and what further revisions to our laws are likely to alter our legal structures to accommodate "thinking" machines. Although statutes and regulations that I believe can spread the benefits of AI technology are addressed further in chapter 9, there are legal areas beyond state and federal statutes that need to address AI, including private law between individuals. The next chapter considers the private and public laws associated with custody and guardianship.

CHAPTER 4

In Robots Parentis: When Robots Have Custody of a Child (or an Adult)

When we think of caregivers, either for our young children or our aging parents, we're typically looking for some combination of tough and soft: tough to handle the stress of providing physical care to someone (who may yell, hit, or break things), and soft to provide compassion and kindness to those same demanding people. We think of strong arms and tender hands. We do not think of machines, despite an earnest effort by Rosie from *The Jetsons* to demonstrate a loving family-care robot in the future.

In some ways, machines have come to be wholly separate from the business of caregiving. Machines pump your blood, measure your vital signs, and stabilize your condition. Machines keep you alive; human beings care for you while the machine does so. While automated machines are capable of performing the rhythmic and monotonous mechanical operations that our bodies do internally for us, machines have thus far been less successful at accommodating the spastic and unpredictable physical responses that the young and unwell externally produce. Robots can pump your blood for years; robots cannot physically control a toddler for five minutes. Those human actions require a higher level of analysis, decision making, and physical dexterity than our machines can currently achieve.

However, during the last 10 to 15 years, therapists, engineers, and people who work with the young and the disabled have become increasingly optimistic that AI technology is slowly catching up to the physical and decision-making requirements that human caregiving demands. This technological progress will become more important as baby boomers demand more assisted-living resources, but the number of qualified and willing nurses and nurses' aides does not increase proportionately. AI will be able to fill in the gap between supply and demand for the caregiving of children and disabled adults. However, AI must be capable not only of telling the difference between a child's harmless tantrum and something more physically dangerous to herself, but also of applying appropriate physical force to both situations without harming the child. AI must be capable of interacting with disabled adults, but also possess the physical skills necessary to pick them up, bathe them, and help them on the toilet.

These AI caregivers don't fit in the established assumption the law makes regarding the care of children: a child is always in someone's custody.[1] During the day, a child might pass from her parents' custody, to her school's custody, to her day care's custody, to her grandfather's custody, back to her parents' custody, as she travels. When that child is left in the custody of an AI nanny, is that chain of custody broken? Does the AI assume custody?

Similarly, many adults become guardians or caretakers for other adults who lack mental capacity—meaning they do not understand the nature and effect of their actions or they do not have the mental ability to understand problems and make decisions[2]—as many developmentally disabled adults do. Additionally, once-capable adults who suffer from Alzheimer's or other forms of dementia are frequently cared for by people acting under authorization from powers of attorney executed before their mental capacity diminished. Those documents typically do not address robot care.

This chapter considers the current laws governing the custody and care of children and adults before looking at the recent development of AI caregivers. With that background, the last section focuses on the legal responsibilities of people and AI when human caregivers use AI caregivers with the people they are responsible for.

Who Has Custody of . . .

Children?

Although teenagers would hate to think of themselves this way, the law assumes that kids are *always* in someone's custody.[3] Typically, they are in the custody of either their parents or the state.[4] A parent's right to custody of their children is rooted in the Constitutional right to liberty, protected under the Fourteenth Amendment.[5] The state's right to custody of children is based on the state's public interest in preserving and promoting the welfare of the child (among other reasons).[6] Included within that broad interest is the right to compel education,[7] to restrict First Amendment expression when such expression materially disrupts school,[8] and the right to detain a child after arrest if there is a serious risk that he or she will commit another crime before his or her court date.[9] Based on these rights, schools act *in loco parentis* (in place of the parents) to search backpacks and ban certain speech and expression.[10] Additionally, juvenile courts act as *parents patriae* (the state as the parent) to intervene to protect the public from crime caused by a child and to protect the child from the consequences of committing future criminal acts.[11]

Although there are many distinctions between the custody of a child in school versus in juvenile detention, free will is not exactly one of them. Although parents can send their kids to private schools or homeschool them, they have to provide education of some sort. It's mandatory by law, so a kid's school attendance is just as compulsory as his or her custody within a juvenile detention center. The circumstances are very different, but the force of law is essentially the same. In both cases, the parent transfers custody to the state. The child is always in someone's custody.

Similarly, parents may temporarily transfer their custody rights, as when a single military parent is deployed and leaves his or her kids with a grandparent or close friend, with the authorization to make decisions in the place of the parent through a power of attorney.[12] Parents essentially do the same thing when they leave their children with a private school, day-care center, or even a babysitter. Obviously, the level of formality changes depending on the sophistication and control exercised by the school, facility, or person. An exclusive private school will require that parents sign extensive documents granting the school authorization for all sorts of child-related issues: emergency health care, discipline, athletics, etc. A day-care facility will typically have a form parents sign that grants day-care staff authorization to perform first aid on the child, to give snacks to the child, and to address other similar issues. A babysitter will typically require payment by the hour, authorization to raid the fridge, and maybe a ride home. In other words, parents place their children into the custody of numerous different parties all the time and grant each certain decision-making rights, but only rarely is a formal power of attorney drafted or signed. But the change in custody, even temporarily, takes place.

With regard to liability, victims of intentional torts are frequently able to recover from the parents of the delinquent child, as the parents are considered vicariously liable for those damages caused by their kids.[13] Similarly, a parent who is negligent in watching his child—such as entrusting a knife to a young child—will be found liable for any damages the child causes because of the parent's negligence.[14] The public purpose of these laws is, among other things, to provide a method to compensate victims[15] (as children are unlikely to have the financial resources to do so) and to encourage parents to watch their children more carefully.[16]

Impaired Adults?

There are typically two ways for one adult to assume long-term legal responsibility for another: a durable power of attorney or a guardianship.[17] In a durable power of attorney, one person (the grantor) grants another person (the grantee, usually a close family member or friend) the right to manage his or her affairs by making important decisions for him or her about healthcare, finances, etc. during a period of disability or incapacity, such as when someone becomes too sick to leave the house or is suffering from dementia.[18] In a guardianship, a court gives one person (the guardian) the duty and power to make personal and/or property decisions for another (the ward). The appointment of a guardian occurs when a judge decides that the ward lacks mental capacity to make decisions.[19] A person who lacks mental capacity is usually someone who is unable to manage property or business affairs because he or she has an impairment in the ability to receive and evaluate information or to make or communicate decisions even with the use of technological assistance.[20]

Among the distinctions between a durable power of attorney and a guardianship is the timing between the two. A durable power of attorney represents a conscious decision by the grantor to give decision-making authority to the grantee. That decision is made before disease or age makes the grantor incapacitated. Largely for that reason, it is typically easier to create and less expensive than a guardianship.[21] A would-be guardian must appeal to a court; the ward either never had mental capacity to properly create a durable power of attorney (such as a developmentally disabled adult) or did not create a durable power of attorney when he or she had capacity (such as an elderly family member suffering from dementia).

Because the guardian is given substantial and often complete authority over the life of the ward, guardianship is seen as a drastic intervention.[22] Whereas the authority in a durable power of attorney represents someone's conscious decision, judges must exercise their best judgment to determine if the ward's loss of self-authority is actually in his or her best interests. The gravity of this decision is part of what prevents guardianship creation from being an easier process.

This legal system for the care and custody of mentally incapable adults is consistent with the long and slow development in England and America of legal safeguards for the self-determination of grantors and wards. By at least the 13th century in England, the king had the responsibility for mentally incapacitated individuals—at the time people used less forgiving words: idiot, lunatic, and natural fool—although there were few ways to ensure that those found to be "natural fools" were actually incapacitated.

Eventually, procedural safeguards were put in place, and in the 17th century a jury of 12 men had to find that the person was without mental capacity, although only those with sufficient funds could really take advantage of this.[23] The jury could find someone mentally incapacitated if the members determined that an individual was, among other descriptions, "of good and sound memory, and by the visitation of God has lost it" or that an individual acts "as a drunkard."[24] In colonial America, family members and community members supported mentally incapacitated adults, but following the Revolutionary War, American courts began to look at English law and cite *parents patriae* as a reason why courts should exercise jurisdiction over those adults in order to protect them and society. Eventually, American states developed elaborate sets of customs, rules, and standards to protect the disabled person.[25]

Similar to liability for children, adults responsible for disabled adults under durable powers of attorney or guardianships can be liable for the actions of the mentally incapacitated adult if the responsible person has been negligent in caring for the impaired adult—that is, if his or her actions were not reasonable or prudent.[26] So, for example, if a man suffering from severe dementia is able to leave his home and, in his confusion, cause a car accident because his guardian did not hire a full-time caretaker, that guardian will be liable if a reasonable guardian should have retained a home nurse. Additionally, depending on a state's laws, responsible adults authorized under durable powers of attorney and guardianships may be criminally liable or subject to civil penalties if they fail to properly exercise their authority.[27]

Parents, grantees, and guardians inevitably provide physically and mentally demanding caregiving to their kids and adult wards. The next section discusses the development of AI and robots that can perform many of those functions, which introduces questions about the rights and obligations of AI and human caregivers. If a human caregiver places an adult with Alzheimer's into the care of an AI caregiver, has the human successfully passed his legal obligation to behave reasonably as a decision maker to the robot? Will that robot be held liable for the actions of that patient? What liability will an AI nanny have for injuries a child sustains while in its care, like breaking a leg while climbing a tree?

Before looking at these questions, it's useful to review how robot and AI caregivers have evolved so far.

AI Caregivers, Machine Aides, and Robonannies

Beginning in the early 1960s, researchers started using computers to remind elderly people about their daily activities, essentially providing a very low level of caregiving to them. Early devices included talking clocks, calendar systems, and similar technologically unsophisticated aids.[28] Efforts continued throughout the late 20th century, with delivery of computerized reminders expanding to the phone,[29] personal digital assistant,[30] and pagers,[31] many of which assisted adults with brain injuries or developmental disabilities as well as the elderly. Although the delivery and sophistication of the reminders changed—advancing from the simple prompts of calendar systems to counseling from telephonic robots[32]— many of the daily activities that the AI (or wanna-be AI) reminded the patients of remained the same: eating, drinking, bathing, medications, money management, etc.[33]

While caregiving continued evolving to include telephones and pagers, the nursing profession began to consider incorporating AI into its caregiving practices, particularly with regard to data analysis.[34] Nursing administration academics in the 1990s noted that "artificial intelligence" had been incorporated into monitoring physiological data and alerting nurses when predetermined "warning" levels had been reached. However, this monitoring function is essentially automation. The computer would trigger an alarm every time a bodily measurement—blood pressure, oxygen, heartbeat, etc.—achieved a designated level. Rather, the suggestion at the time that AI could be used to manage quality and ensure better outcomes is closer to actual AI, as it relied on programs or machines to analyze patient service data and make decisions themselves.[35]

In 1998, researchers from the University of Michigan, the University of Pittsburgh, and Carnegie Mellon University joined together to form the "Nursebot Project," whose goal was to develop mobile robotic assistants for elderly people living in their homes, particularly those with cognitive impairment.[36] Among these researchers was Sebastian Thrun, before he won the DARPA Challenge and took Google Car to the streets. Earlier in his career, while at Carnegie Mellon and Stanford University, Thrun was an enthusiastic and outspoken advocate for introducing AI into nursing homes and assisted-living facilities, permitting the robots to interact with patients and other adults needing care.[37]

The Nursebot Project was mindful of the need to create AI that would be flexible, in order to respond appropriately to the actions of its patients. For example, an AI that observes a patient with urinary incontinence problems use the bathroom at 10:40 a.m. should know to reschedule the planned bathroom reminder originally set for 11 a.m. Similarly, if the patient's favorite TV show is on at 1:30 p.m., that same AI should know to issue the rescheduled reminder earlier—perhaps at 1:20—noting the reason why going to the bathroom at that time would be useful: to avoid missing that show (e.g., "Mrs. Adama, if you use the toilet now, you will not miss your favorite episode of *Battlestar Galactica*, the one with the Cylons"). Additionally, the developed AI will contain sensors and gather information about the patient's activities and habits, making inferences during the day. So if Mrs. Adama goes to the kitchen around her normal dinner time, that may indicate that she is beginning dinner. In following a patient around throughout the day, the AI will able to assist that person in navigating her environments, which is important for assisted living facilities, where nursing staff spend a large percentage of their working days escorting patients from one location to another.[38]

In 2001, the Nursebot Project tested "Pearl," a robotic nurse, at the Longwood Retirement Community in Oakmont, Pennsylvania.[39] Pearl escorted patients through the facility and reminded them to take their medications. "The entire place was really into it," Thrun said,[40] Pearl also provided information of interest to the patients, such as weather reports and television schedules.[41] Although researchers felt that Pearl was a success,[42] there are no future plans for her.[43] Nonetheless, many people in robotics and health care believe that AI and robots will become integrated into nursing and caregiving in the near future. Linda Hollinger-Smith, vice president of the Mather Lifeways Institute on Aging, believes robotic assistive devices will be a part of long-term care in the next 15–20 years. Debi Sampsel, former executive director of the Nursing Institute of West Central Ohio and current Chief Officer of Innovation and Entrepreneurship at the University of Cincinnati College of Nursing, believes that robots offer a solution to today's nursing shortages.[44] Thrun believes that robots will incorporate more and more AI functions, providing greater benefits to the elderly.[45] Some functions may be coming to homes sooner rather than later. Siasun Robot & Automation Co., Ltd. in China believes it can bring a family nanny robot—that interacts with elderly family members and monitors their vital statistics—to the market by 2015.[46]

And while much of the development thus far has been on the cognitive side of caregiving—providing reminders, monitoring activities—researchers are directing more time and energy to the physical side of caregiving: lifting patients, holding objects, etc. The U.S. Department of Defense has granted a contract to RE2, Inc. to develop a multitasking robotic nursing assistant (RNA) that will be able to lift combat-wounded soldiers out of bed and deliver supplies.[47] Jessica Pedersen, the former chief operating officer of RE2, Inc., would like RNAs eventually to navigate hospitals and nursing homes.[48] Significantly, researchers at the Healthcare Robotics Laboratory at Georgia Institute of Technology have worked on AI that will autonomously wash patients.[49] The researchers have permitted the AI, which they have named "Cody," to clean limited patient body parts, such as the arms and legs, but hope that further development will permit Cody to effectively and safely clean other areas of the body.

Those two qualities—cognitive ability to oversee persons and physical capability to directly and safely interact with people—will also be necessary for AI to act as nannies to children. With an eye toward AI that can perform those functions for parents, scholars and ethicists are already debating whether this is a good thing. As early as 2008, ethicists like Noel Sharkey have warned that "the use of service robots poses unanticipated risks and ethical problems ... in the care of children."[50] Sharkey worries about a parent leaving a child in "the safe hands of a future robot caregiver almost exclusively," noting that studies have shown social dysfunction occurs when infant animals develop attachments only to inanimate surrogates.[51] In response to these concerns, Joanna Bryson notes that robot care is more likely to complement human care, not replace it. Children may bond with their AI caregivers, but that could reinforce their sense of self-worth.[52] And although truly effective AI babysitters and nannies will not be on the market anytime soon, manufacturers are already performing the research and development necessary to produce models.

For example, NEC Corporation in Japan has created the PaPeRo, a prototype of a "communication robot" that tells jokes, gives quizzes, and uses radio-frequency identification chips to track kids.[53] Researchers at NEC have played with the PaPeRo since 1997, but currently only demonstrate it at tech shows to advertise its development. The company does not sell a model.[54] However, although the prototype does not have the physical capabilities of the RNAs commissioned by the U.S. Department of Defense, it can recognize faces, speak, autonomously wander through rooms, and travel to its designated charging area when its battery drops to a certain level.[55] This level of AI is currently inappropriate for a babysitter, but would permit PaPeRo to act as an AI companion for children of a certain age, entertaining a young child while a parent or guardian is in another part of the house or possibly next door. In 2008, Japanese retailer Aeon began offering robot babysitting services in its Fukuoka Lucle mall. Children wear special badges with QR codes so the robot can identify each child.[56]

More recently, Maja Matarić, a computer science professor at the University of Southern California, has reported that her robot, Bandit, is able to help autistic children, effectively interacting with them, collecting information about a child's behavior, and even using that to predict future behavior from the child, like when he or she is about to stop exercising.[57] Similarly, some schools in the United States are incorporating robots into the classroom that assist autistic students.[58] Other research centers are also working with robots as a learning tool for children. Elizabeth Kazakoff is a researcher with the DevTech Research Group at Tufts University, which is in the process of developing a robotics building kit to assist kids with specific developmental needs. She is optimistic about the ability of robots to assist children and families. "I think robots are great educational tools," she says. "There is fairly new research on the concept of care-receiving robots—where young children 'teach' robots in order to learn the concepts themselves. Teaching the care-receiving robots, in turn, facilitates their own learning."[59]

Kazakoff also notes that researchers are exploring robot teaching assistants. The South Korean government is in the process of introducing hundreds of AI teaching assistants to preschool classes. The AI assistant, which looks like a small dog, marks student attendance and uses a face-recognition program to ask children about their mood. The researchers behind this technology admit, though, that AI can only supplement a human teacher at this time. "Due to the limits on the current robotic technologies, robots cannot completely supplant human teachers in the educational field," explains Mun-Taek Choi, a senior research engineer at the Korea Institute of Science and Technology (KIST).[60] "Robots are very helpful to enhance the concentration capability of children in class," notes Bum-Jae You, the head of the Cognitive Robotics Center at KIST, but "[t]eaching is probably the most challenging role for artificial intelligence . . . It requires a real fundamental leap in ability before we can get there."[61]

Arguably, Asimo, Honda's humanoid robot, is the most publicized robot currently under development to perform physical and cognitive human functions. Honda researchers have worked on robots since 1986, and Asimo, the most recent model, is able to pick up a glass bottle, twist off the cap, and pour liquid into a soft paper cup without crushing it. Cognitively, Honda claims that Asimo can recognize the voices of multiple people speaking simultaneously. This combination of physical and cognitive is impressive and promising for a potential AI caregiver, but Honda admits that there are no practical uses for Asimo at this time.[62]

The AI functionality of these models is fairly limited—small vocabulary, few activities to engage in with kids, no physical capacity to restrain or interact with children, etc.—but it suggests that more advanced AI is on its way to the market, ready to look after our kids and possibly our aging parents too.

What Happens When a Robot Babysits Your Toddler or Cares for Your Grandmother?

The AI described above is not likely to bring a fully functional AI caregiver to Best Buy or Amazon anytime in the near future. But that hasn't stopped some parents from daydreaming about robot babysitters that can:

- Stay home for 10 minutes to watch the napping baby while they drive to school to pick up their oldest child;[63]
- Watch the kids so they can run to the store to pick up more milk without getting four kids into a minivan;[64]
- Monitor their children constantly, providing for the child's needs as they arise and giving feedback to the parents;[65] and
- Babysit their son while they go out for a night on the rings of Saturn, making sure he goes to bed without cookies, milk, or jokes.[66]

Similarly, professionals who work in adult caregiving hope that AI caregivers can assist human nurses and caretakers to bathe patients,[67] monitor patients,[68] and provide social interaction with patients,[69] among other functions. These forms of AI will be in addition to autonomous vehicles that will be able to shuttle children and mentally incapacitated adults to other locations for classes, therapy, and other activities, while also monitoring those riders.

With the exception of Benjamin McFadden's parents traveling to Saturn, these scenarios present serious questions about how the law will treat parents who leave their kids in the care of a robot, human caregivers who place their adult charges in the care of AI caregivers, and the AI placed in the position of unsupervised caregiver. Let's deal with these actors one at a time: the adult caregivers, the parents, and the AI.

What happens when a human caregiver leaves a patient suffering from dementia with an AI caregiver and there's an accident?

When a human caregiver leaves their wards with an AI caregiver, the first question will be "Was that reasonable?" Negligence, as discussed in chapter 2, will be a common claim against human operators of AI. Initially, human caregivers will likely have to answer a lot of questions to show that leaving incapacitated adults with a robot was reasonable: "What does the ward suffer from? How long were you gone? How far away did you go? What level of functionality does the ward have? What type of AI caregiver was it?"

These questions will be very important at first. Until AI caregivers become more common place and are considered a known technology, judges, juries, and the law will be skeptical of them. So when AI caregivers are first available, shorter trips to the store will be viewed as reasonable, but robots caring for mentally incapacitated adults all workday will not. Leaving calm mentally incapacitated adults with AI will be reasonable, but leaving unstable mentally incapacitated adults with AI will not. Letting AI watch a sleeping mentally incapacitated adult will be reasonable, but letting AI watch the same adult while she works in a kitchen will not. Some AI caregivers will have better reputations than others and more research supporting their safety. Adult caregivers that place their wards in the care of properly vetted AI for reasonable periods of time in the right circumstances will be considered reasonable, regardless of whether or not the mentally incapacitated adults get a bruise while in the care of the AI. So if a guardian reasonably leaves her sister, who has Down syndrome, with an AI caregiver while she goes down the street for some milk and the sister suffers a concussion, the guardian should not be liable. She acted responsibly.

Similarly, a guardian caring for her grandmother, who suffers from dementia, should not be liable if, after placing grandmother in the care of an appropriate AI caregiver, grandmother runs confused out into the street and causes a car accident. The guardian acted responsibly, but the AI was nonetheless unable to prevent the grandmother from causing damages.

On the other hand, if a guardian places grandmother in an autonomous car that is not properly designed to contain a dementia patient and grandmother exits the car and causes a traffic accident, the guardian would be liable for negligently placing grandmother with an inappropriate form of AI. The design of the AI and the mentally incapacitated adult's level of function (high or low) are important considerations for guardians. So before using autonomous cars to transport incapacitated adults (or children) by themselves, guardians must make sure that the vehicle: is able to ensure that the passenger will only be released in the presence of another responsible adult; can redirect itself to appropriate care if the passenger appears to suffer an injury (stroke, seizure, concussion, etc.); can properly restrain the passenger; etc.

AI caregivers should give grantors pause as much as grantees and guardians. As AI caregivers get closer to the market, attorneys may want to discuss with clients the idea of including language in their durable powers of attorney that addresses AI caregivers. Do the clients hate the idea of receiving care from a robot? Do they like the idea? Will they be okay with AI caregivers once certain safety standards have been satisfied? If aging clients have concerns about AI caregivers, they should consider addressing those concerns in their durable power of attorney documents. (Similarly, in the not-too-distant future, parents should think about AI babysitters when drafting power of attorney documents to permit friends or family to make decisions about their kids while they are unavailable.)

What happens when parents leave children with AI and the children cause property damage?

Like guardians and people authorized under a durable power of attorney, parents can be liable for injuries and damages that occur due to their negligence while caring for their kids. But parents can also be liable for the intentional torts of their children. So if a kid is in his parents' custody and breaks windows while throwing rocks at them, the parents will be liable.

So who is liable if the same kid throws rocks and breaks windows while in the care of an AI babysitter? Assuming the AI operated properly and there are no design or production flaws in the AI, which would mean the manufacturer or designer is liable, I suspect that laws will be drafted to address that situation in one of two ways:

1. *The parents will continue to be liable.* In some ways, this speaks to our sense of fairness. "They're your kids. You're responsible." And I think that some states will legislate it this way. But if the parents have behaved reasonably by putting their child in the care of a properly designed and manufactured AI babysitter, I do not believe it is reasonable for the parents to be liable. They appropriately tried to ensure their child's safe actions. Liability should follow fault; there was none here.
2. *The AI will be liable.* Consistent with the discussion of AI liability in chapter 2, I think it is more appropriate for the AI babysitter to be liable when the child commits an intentional tort. The AI will be designed to reasonably take care of the child and prevent it from doing harm to itself and to others. The AI should be liable and should pay, using either an insurance policy taken out on the AI or the proceeds from a government-organized AI reserve fund.[70]

So, for example, if Dad runs down the street to get batteries while his four-year-old daughter plays in the yard under the care of an AI babysitter, he should not be liable if she gets irrationally annoyed at the neighbor's son playing nearby and rides her bike into him, breaking his arm. He left her with the AI, which existed in that moment for the sole purpose of protecting the best interests of the child. The AI failed and should be responsible.

To use an example that could occur now, suppose the Aeon babysitting robot at Fukuoka Lucle mall in Japan is responsibly watching a child, but the child still manages to run out of the child-care area and trip an elderly woman. Should the parent be liable for that kid's intentional tort? The parent reasonably left the child with supervised care, meaning the AI should be responsible. Having said that, the AI cannot be liable under our current laws, even when it is at fault.

Chances are, the store would be liable, likely for negligent use of the AI. I don't find this to be an unacceptable outcome, but I dislike the reason. If the AI was *responsibly* caring for the child, the store was not negligent. Rather, the AI was acting as an agent for the store, basically a robot employee, and the law should treat the robot as a human employee in that situation. So the store would be liable under the theory of *respondeat superior* because the elderly woman was injured due to the actions of the AI as a babysitter while it worked as a babysitter in the store.

What happens when a child or mentally incapacitated adult is injured while under the care of AI?

When a guardian or parent places a mentally incapacitated adult or a child in the care of AI and the adult or child is hurt, the AI may or may not be liable. If the AI was improperly designed then the developer is liable. If the design was appropriate, but a flawed AI was produced, then the manufacturer is liable. If the AI was properly designed and manufactured but was damaged during shipping, the carrier service is liable. If the AI was properly designed and manufactured, with no user error on the part of the parent or guardian, then the AI itself is liable. These liability issues will be addressed, I believe, in the specific-AI legislation and Uniform AI Acts that states will pass to address liability for AI.

Closing Thoughts on Robonannies and AI Caregivers

The laws governing custody of children, parent liability, decision-making authority under guardianships and durable powers of attorney, and grantee/guardian liability will need to be revised to account for AI caring for and making decisions about those kids and mentally incapacitated adults. AI caregivers will make decisions about going outside, what to make for lunch, other people entering the house, etc. They will need to be able to perform basic first aid, give baths, and restrain upset children and adults. In short, AI will have to be physically soft enough to put a baby into a crib and tough enough to handle a thrashing adult, while also being programmed to monitor the kids and mentally incapacitated adults in their care and make cognitive decisions about their needs. Are they hungry? Do they need more exercise? Do they need social interaction?

This combination of complicated physical ability with cognitive analysis will also be required of AI that works at farms and in industrial plants. Such robots will make industry attractive where it currently is not—commercial and residential zones—while also challenging assumptions we have about farming and its impact on a community's character. How zoning ordinances can address those changes is discussed in the next chapter.

CHAPTER 5

RIMBY (Robots in My Backyard)

Thoughts of widespread AI conjure images of autonomous cars and robot housekeepers, not cities and town. Until recently, the last serious conversation I had about AI and cities took place when I was 10 and was about Metroplex, Trypticon, and whether the living Autobot City could defeat the sentient Decepticon command base. Few people consider the impact AI will have on the most basic laws and regulations governing our communities, municipal and zoning ordinances. Local ordinances dictate where we can build houses, open businesses, walk the dog, park our cars, and many other mundane aspects of our day-to-day lives. Towns, city councils, and boards of selectmen will have to amend local ordinances and accompanying regulations to reflect the introduction of AI to the local economy. In much the same way that state and federal law is based on the premise that only people make decisions, local ordinances are based on the premise that only people make decisions *and* use town roads and properties. That increasingly will be an incorrect assumption.

When autonomous cars are able to drop their owners off at work and return home, parking fares in downtown areas will need to be revised because few people will park there. When entrepreneurs can rely on AI workers to help with home businesses, zoning ordinances that control the size of home businesses in residential areas by limiting the number of employees will be ineffective. When children arrive at school unattended but for their parents' Google Car, school boards will have to revisit pickup and drop-off rules. When AI store and factory workers are able to operate around the clock, which is already happening in some places,[1] towns that never bothered to address noise or activity limits during evening hours will have to revise their ordinances to protect nearby residences from stores and plants that start operating 24 hours each day, seven days a week.

This chapter looks at local ordinances—primarily municipal and zoning—and suggests how AI will force us to change them. The two primary areas of concern are appropriate control over land use and loss of revenues. As discussed below, the sanctioning of land use controls by the U.S. Supreme Court in 1926[2] has permitted towns and cities to assume vast control over their development and composition. Many towns and cities have revitalized their local real estate markets and economies by using proper city planning and zoning controls. However, AI potentially allows clever landowners to circumvent zoning restrictions, permitting industry and commercial activity in areas where town residents decided only residences, schools, and churches should exist.

Additionally, cities and towns depend on local ordinances for revenue. This revenue is threatened by the growth of AI. An entrepreneur without AI is limited to hiring two employees for her home brewing business; this effectively limits the amount she can produce and sell from her home. If she wants to grow beyond two employees, she'll need space that is zoned for commercial use. That same entrepreneur is able to grow a substantially bigger business in her home when she can use AI workers, which are not limited under the local ordinance. By keeping the business at home, her property taxes will continue to be a residential rate, not a commercial rate, which is typically higher. Similarly, towns and cities that depend on parking fees for their revenues will lose some of that as more people send their autonomous cars home for others to use, rather than pay to leave them unused in a parking lot.

In order to demonstrate changes necessary to local ordinances, I trace the history of robots in industry, discuss the current development of AI industrial workers, and then show how civic leaders might need to change their local ordinances. Before doing so, a more complete understanding of local ordinances and their history is necessary.

Local Ordinances—What Do They Do?

Although related, municipal ordinances, which govern behavior in a municipality, and zoning ordinances, which govern use of land in a municipality, are distinct forms of law with separate histories.

Municipal Ordinances

In the list of laws that trump one another, municipal ordinances are essentially at the bottom. They are subordinate to the U.S. Constitution, federal laws, treaties, presidential executive orders, state constitutions, and state laws. Indeed, the municipalities themselves only exist through state charters or authorization.[3] Given this status, many believe that towns and cities are essentially powerless,[4] as they have no legally enforceable right to exist, to control local resources, or to regulate local territory.[5] At the same time, practically speaking, municipalities have a great deal of power, as they typically make major policy decisions—safety, public health, recreation, etc.—without significant external supervision.[6] This contradictory authority—weak in theory but strong in practice—reflects the historical development of municipalities and their laws.

Many English towns in the medieval period formed as groups of people seeking protection from outside forces.[7] In particular, merchants banded together to enhance their economic power. People and real estate in the Middle Ages were subject to numerous competing jurisdictions—the nobility, the king, the sheriff, etc. By banding together, the merchants of a new town could negotiate with those competing forces and obtain a certain amount of autonomy by obtaining a charter as a municipal corporation. Because commercial enterprises also need charters, for centuries English law treated public corporations (cities and towns) the same as private corporations (businesses like the East India Tea Company). Under English corporate law, this permitted municipalities to make rules and laws governing activity within the town, particularly commercial activity. A town association, the municipality's governing council, protected workers from exploitation, regulated labor conditions and prices, punished fraud, and asserted the town's interests against neighboring competitors.[8]

This model existed without great challenge for several centuries, as the corporate charters that specified the local municipality protected its autonomy from intrusion by the king. Not that the king didn't try. As early as the 1200s, kings asserted that the monarchy could revoke charters for wrongdoing, although without much success. It wasn't until the late 17th century that the monarchy scored a significant victory against the independence of towns and cities. Alleging that the City of London imposed unlawful tolls and published "malicious and seditious libel,"[9] the king won a court case that established royal control of cities, at least for a short period of time. I say a short period because the Glorious Revolution removed the Stuart dynasty from the throne, which resulted in the reversal of the London case. This established that municipal charters were protected from *royal* control but not from *parliamentary* control. This retained the idea that towns and cities did not have true autonomy, only autonomy at the will of the central government.[10]

With this heritage behind them, American colonists first created towns in much the same way that medieval merchants did, as associations to protect their common interests. However, municipal corporations were not common, as there were only about 20 incorporated cities in pre-Revolution America. This can be partly explained by the fact that colonial America did not have a strong central government to negotiate or contend with. The king and parliament were across the Atlantic Ocean. There was little need for charters to protect cities from a central authority. Even as colonial (later state) governments developed by the late 18th century, those governing bodies were composed of representatives from towns, who had instructions to protect the towns' interests.[11]

However, as American political and legal life after the American Revolution began to revolve increasingly around state legislatures, states became dominant over municipalities, interfering with the ordinances and affairs of cities and towns at will.[12] Partly because of the way American state governments grew and partly because of lingering English ideas about central government's power over municipal corporations, the conventional legal theory that developed determined that states enjoy complete power over local governments.[13]

In response to concerns regarding the limited scope and capacity of municipal governments in the 19th century, many states amended their constitutions, permitting "home rule."[14] Under home rule, state constitutions and legislation granted greater powers to chartered municipalities, protecting them from state intrusion when acting in "local affairs." Today, ordinances passed by municipalities are used to enact a permanent rule of conduct,[15] which can govern a broad array of issues, including the provision of basic public services like education;[16] use of streets and town parks;[17] procedures for the town or city to conduct its elections, business transactions, tax assessment and collection, and acquisition of new real property;[18] and creation of city courts.[19]

That is not to say that states do not continue to exercise oversight of municipalities; controversies regarding local property taxes and school funding have been resolved by state courts and in state legislatures.[20] But overall, state legislatures honor the authority of municipal ordinances by declining to pass laws that would preempt those ordinances in critical local affairs.[21]

Zoning Ordinances

The development of home rule and stronger municipal ordinances contributed to the creation of zoning ordinances in the early 20th century. Zoning ordinances are local regulations designed to prevent the harmful effects of land uses (smelly pig farming, polluting factories, etc.) from impacting neighboring properties. In general, the idea is to separate land uses, so different ones will not harm one another. Houses will be in one area of town, stores in another, public buildings in yet another, and industrial land in another, far-off corner.[22]

This was not an innovation of *SimCity*, as some mistakenly believe. Rather, zoning ordinances were a direct response to the Industrial Revolution, "the main creative force of the 19th century [which] produced the most degraded human environment the world had yet seen."[23] Zoning is based on the ideas of Ebenezer Howard, who was appalled by the chaos of industrial London and wrote a book in 1898, *Tomorrow: A Peaceful Path to Real Reform*, that managed to gain quite a following despite one of the most lackluster titles in the history of influential books. Howard's main idea—resettling a large portion of the urban population to new towns with ample space to live healthy and useful lives—was not of interest to American city planners in the early 20th century. However, they liked the principles underlying his main idea:

1. Separation of uses;
2. Protection of the single-family home;
3. Low-rise development; and
4. Medium-density of population.[24]

Despite the preference these principles gave to small or medium-sized towns, large American cities were among the first municipalities to embrace zoning control of land. Los Angeles enacted a zoning ordinance in 1909, restricting industry to specified districts, away from residential areas. New York City enacted the first comprehensive zoning program in 1916, not only separating uses but also restricting the height and bulk of buildings. This program came partly in response to haphazard growth of tall skyscrapers (which cut out light and air for many residents), but also because garment manufacturers had started to move into spaces close to posh Fifth Avenue stores. Merchants thought the riffraff was bad for business. City leaders became convinced that zoning was essential for a great metropolis.[25]

Zoning spread rapidly after New York adopted its ordinance (presumably if zoning was good enough to make NYC a little nicer, bucolic America must have felt it was good enough to keep dirty factories from ruining the good part of town), particularly after the U.S. Commerce Department adopted the Standard State Enabling Act in 1922, which made it easier for states to authorize towns to implement zoning ordinances and guide them through how those ordinances would function.[26] By 1926, 420 American municipalities with a total population of 27 million people were zoned. All but five states had passed zoning enabling legislation.[27]

However, in the mid-1920s, state courts began declaring zoning ordinances unconstitutional.[28] The issue was property rights: Does a town have the right to restrict a property owner's use of his or her property? This, of course, is exactly what zoning does. Towns inform owners, "We have decided that your property is in an area where we want residences only. We don't care that you bought it hoping to use it as a widget factory." Only owners who were already using their properties before the ordinance was passed or amended may continue to use land in a prohibited way, which is known as a grandfathered right. If the wanna-be widget-maker had opened his factory a year before the ordinance was adopted instead of a year after, he would have been fine.

Many (particularly real estate developers) felt preventing property owners from doing anything they wanted on their properties violated § 1 of the Fourteenth Amendment, depriving those owners of liberty and property without due process of law.[29] This argument was put to the test in *Village of Euclid v. Ambler Realty Co.*, the 1926 Supreme Court case that stated definitively that zoning ordinances as a concept are constitutional, although specific provisions may become "arbitrary and unreasonable."[30] In *Euclid*, a real estate company owned 68 vacant acres that it intended to develop for industrial use. However, the Village of Euclid adopted a zoning ordinance before such development could occur and essentially permitted only residences on the tract of land. This reduced the property's value from $10,000 per acre to $2,500 per acre. Ambler Realty asserted that its lost value was an unconstitutional taking: through its ordinance, the village had taken $7,500 per acre of value from Ambler without compensation.[31]

However, the Supreme Court rejected Ambler's argument, upholding Euclid's zoning ordinance. In doing so, it noted that the village was properly exercising its police power to separate industrial and commercial establishments from residential areas. A municipality may create a general plan for uses within its territory to promote the health and safety of the community.[32]

Based on this decision, cities and towns have almost universally adopted zoning ordinances, with as many as 97 percent of municipalities in the United States using ordinances by the end of the 20th century.[33] Although smaller communities frequently use fairly simple ordinances—sometimes using only one or two zoning districts to regulate the entire town—many municipalities have enacted increasingly complicated zoning ordinances in an effort to perfectly plan how they want their communities to grow. Zoning ordinances still separate industrial and commercial uses from residential uses—and certain types of residential uses from each other, like apartment complexes from large single-family homes[34]—but also frequently address uses in very detailed fashion. At the time of the *Euclid* case, zoning ordinances might list fairly specific permitted uses; Euclid's ordinance, in fact, noted permitted uses like "stations for gasoline and oil (not exceeding 1,000 gallons storage)," "sanitariums," and "stable and wagon sheds (not exceeding five horses, wagons, or motor trucks)."[35]

But towns today, even small towns, are more likely than they were 80 years ago to address things like:

- Home businesses, that is, businesses or professional offices operated out of someone's residence;
- The amount of parking required for any commercial or industrial space;
- Monthly vehicular trips permitted to any new industrial development; and
- Required loading space and dumpster space for commercial developments.

These overlap with other municipal regulations but fit within zoning as they impact how land is used and address questions that nearby property owners may ask. "Will the birdhouse factory bring hundreds of truck trips past my business each week?" "Will delivery trucks stop in front of the new restaurant for an hour, impeding traffic to my store?" "How many people will be working in Frank's garage now that he's professionally detailing cars there?"

There are also common zoning ordinance provisions addressing special review for telecommunications towers, development within aquifer areas, or special zoning districts to promote specific uses already in place (like a stadium or theater). But inherent in most of these uses is the assumption that it is people who are using the land. That will not always be the case anymore as AI begins using properties for people, but without the need for people to be present on the property making decisions.

Industrial Robots and AI Industrial Workers

Nowhere is this more true than in industrial AI. The industry sector, which originally prompted the development of zoning ordinances in the wake of the Industrial Revolution, will incorporate AI as a way to cut costs but will inadvertently disrupt many existing municipal and zoning ordinances. Before discussing how, let's look at the development of robots and AI in industrial production.

The Idea of Industrial Robots

In general, an "industrial robot" is an automatically controlled, reprogrammable, multipurpose, manipulative machine, which can either be fixed in place or mobile for use in industrial settings.[36] The main applications for industrial robots are:

1. Processing (the robot handles a tool);
2. Materials handling (the robot, equipped with a gripper, handles the components of products);
3. Assembly (the robot puts product components together by insertion); and
4. Testing/handling (the robot checks ready-made or incoming goods in various ways).[37]

Not surprisingly, these are the same areas of industrial activity where AI promises to have the fastest impact.

The idea of an "automatic tool," though, has deep roots in Western Civilization. In 322 BCE, Aristotle wrote: "If every tool, when ordered, or even of its own accord, could do the work that befits it, then there would be no need either of apprentices for the master workers or of slaves for the lords." In 1495, Leonardo da Vinci designed a mechanical device that looked like an armored knight and had internal mechanisms designed to move the device around like a real person.[38] The word "robot" was created in 1921 by the Czech playwright Karel Čapek, who based it on the Slavic term *robota*, which means heavy monotonous work or slave labor (drudgery). His play *R.U.R.* features worker robots that do the jobs of human beings.[39] Isaac Asimov, no stranger to strange ideas about robots, predicted in 1941 that a powerful robotics industry would develop.[40]

Early Development of Industrial Robots

In fact, it wasn't long after that that robotics became an industry. Actual industrial robots began appearing in manufacturing plants in 1961, when Unimation, Inc. installed the first industrial robot, a hydraulic arm, at General Motors' Ternstedt plant in Trenton, New Jersey. Obeying a step-by-step program from a magnetic drum, the robot stacked hot pieces of die-cast metal,[41] a job that was very unpopular with plant workers.[42]

Unimation was formed by Joseph Engelberger, who has since been called the "Father of Robotics." Fascinated by science fiction and the robot, Engelberger pursued physics at Columbia University and became an engineer in the aerospace industry. By chance, he met George Devol Jr. at a cocktail party in 1956.[43] Two years earlier, Devol had filed a patent application for "Programmed Article Transfer," which made possible a general-purpose, automatically operating robot.[44] Engelberger later licensed Devol's patent to create Unimation. That first robot in 1961 was sold at a loss, and Unimation did not post a profit for a number of years. But the company eventually became incredibly successful, with $70 million in sales by 1983.[45]

Those sales figures were possible because between 1961 and 1983 industrial manufacturing began to embrace robotics, sometimes in surprising ways. The image of robotic arms slowly dominating automotive factories is fixed in popular history, but what about wheelbarrows? A 1964 Norwegian labor shortage forced wheelbarrow manufacturer Trallfa to develop a robot that could spray-paint its wheelbarrows.[46] The spray-painting department at Trallfa had undesirable working conditions (fumes, etc.), making recruitment difficult. The spray-painting robots Trallfa developed could perform continuous movements and were easy to program. Although Trallfa intended to limit their use to internal production of wheelbarrows, the design was so successful that Trallfa began selling the robots to other manufacturers.[47]

The magnetic drum that Unimation used to provide direction to its hydraulic arm in 1961 seems hopelessly antiquated today; it is like comparing a vinyl record to an iPod. Similarly, it would be impossible for robots to perform much of the industrial work they do today operating exclusively on hydraulic power. Industrial robotics design began to move toward modern computers in 1971, when Cincinnati Milacron Inc. began marketing The Tomorrow Tool (with the catchy, though sadly Schwarzenegger-free, title "T3").[48] This was the first computer-controlled industrial robot.[49] In 1974, the Sweden-based Asea Group, which became Asea Brown Boveri Ltd. (ABB) shortly thereafter,[50] introduced the first all-electric, computer-controlled robot, the IRB 6, which was designed for industrial grinding.[51] However, it proved to be an adaptable technology, as the first purchaser was a German firm, Magnussons, that used it to wax and polish steel tubes.[52] Magnussons had a small workforce of 20 employees. Using the IRB 6, Magnussons was one of the first manufacturers in the world to operate an unmanned factory, 24 hours a day for seven days a week.[53]

During this time, painting and welding robots grew into the most common industrial robots. In 1976, British agricultural machinery manufacturer Ransome, Sims, and Jeffries modified Trallfa's spray painting design for use in arc welding.[54] Contributing to the growth of painting and welding robots in industrial settings was the fact that Unimation's robots were adaptable for spot welding.[55] GM began purchasing Unimation welding robots in 1969.[56]

The relationship between GM and Unimation assisted the development of the American robots market. The "standard arm" design that we associated so strongly with industrial manufacturing (think long and mobile arms flipping backwards to grab hoods before flipping back forward to install them in cars) originated as a project from the Stanford Research Institute, which, as SRI International, did the research that led to Siri and the da Vinci surgical robot system. In 1977, Unimation purchased the standard arm design with backing from GM.[57] In 1978, the company released the Puma (Programmable Universal Machine for Assembly),[58] which became instrumental in the development of surgical robots.[59] The Puma eventually gained widespread acceptance in automotive factories.[60]

Worldwide Spread of Robots in Factories

Although much of the initial development of industrial robots occurred in the United States, Japan proved much more eager to adopt robots in manufacturing. In 1980, Japan became the dominant user of manufacturing robots. Between 1978 and 1990, Japanese industrial robot production grew by a factor of 25. The worldwide industry underwent a period of major consolidation between 1984 and 1990, as companies devoured each other. Partly because of the lucrative Japanese market for industrial robots, only a small number of non-Japanese firms survived.[61] By 1995, the Japanese robot market represented 70 percent of the worldwide market.[62]

However, certain European manufacturers have established dominance in the field as well. By 2005, the two largest robot manufacturers in the world were Japanese; the next two were European.[63] One of those was ABB, which has established a dominant market position, partly because of the firm's strategic acquisitions. In 1985, it purchased Trallfa. In 1990, it purchased Cincinnati Milacron's robot division,[64] which was one of the largest robot manufacturers in the United States in the 1980s and represented as much as 32 percent of the American market.[65] The European production of industrial robots has grown as the number of robots in European factories has increased. In the last few years, Europe's stock of working industrial robots has pulled essentially even with Japan's numbers, while Americans use less than half of either.[66]

The increased presence of industrial robots in production has been sharp. In 1973, there were 3,000 robots used in industrial production; by 2011, there were 1.1 million.[67] In 1961, the hydraulic Unimation machine installed in a GM factory could perform one function: stack pieces of hot die-cast metal. Fifty years later, industrial robots weld, paint, handle and assemble components, operate in industrial clean rooms, process products, grind, wax, and perform numerous other tasks.[68] The driving force behind the expansion of robots in industrial production is the reduction of cost. In Japan, between 1970 and 1985—when the Japanese industrial sector embraced robots—labor costs increased by 300 percent while robot prices declined by nearly 50 percent. In the mid-1990s, wages increased worldwide more rapidly than robot prices.[69] That trend has continued. Between 1990 and 2005, the cost of an industrial robot has decreased by 46 percent without factoring in labor costs, but that cost has decreased by 90 percent when factoring in the rising cost of labor.[70]

AI Industrial Workers

There is no reason to believe that these numbers will suddenly start to reverse themselves, particularly as aging populations in many industrial countries make *finding* workers difficult—never mind *paying* them. Rather, would-be employers up and down the chain of production will continue to seek technologies that permit them to reduce their reliance on expensive workers. AI industrial workers in use or in development now do just that. (By the way, for the purposes of this section, the term "industrial robot" will refer to robots that work in the manufacturing sector, but also those that work in the "primary" sector collecting natural resources, in order to provide a broad view of AI in production.)

As a sample, consider the following fields and functions that are exploring the adoption of AI, or have already started using AI industrial workers:

- *Agriculture:* Given its affinity for robots and aging population, Japanese companies and researchers have naturally begun looking at ways to introduce AI to the labor-intensive task of sowing and harvesting fields. Japan's National Agricultural Research Centre has developed a robotic rice planter that is capable of moving autonomously through paddy fields.[71] It uses GPS to map the field and a computer to program the planter's path through the flimsy and difficult field terrain based on the GPS data.[72] Additionally, researchers at Miyazaki University and Kyoto University are developing autonomous strawberry harvesters. The current prototype features strawberries growing in raised beds, with rails between them. AI workers travel on the rails and use sensors to determine if individual berries are ripe. If they are, the AI picks them with a gentle suction before transferring them to a conveyor belt or basket. So far, they are not as fast as human workers (although they can work 24 hours a day), and the cumbersome mechanism is not popular with strawberry farmers. But the farmers also admit they might need a system like this someday.[73]

  In related technology, the Massachusetts Institute of Technology has developed an experimental tomato greenhouse that is managed entirely by small robots. Each plant is monitored by sensors. If the sensors identify a plant as dry, a robot is dispatched to water it. When a tomato is identified as ripe, the robot is able to locate it on the vine and pick it with a mechanical arm.[74] In addition to picking and watering the plants, the long-term goals for the robots in this greenhouse project, titled the Distributed Robotic Garden, include weeding and cleaning (removing dead leaves).[75] Assuming researchers are able to develop AI that can effectively identify and remove dead leaves, those robots will have greener thumbs than most people who keep plants in their apartments.

- *Mining:* Skilled labor shortages and metal ore accessibility have driven the mining industry to begin embracing AI. Large, easily obtained deposits have all but disappeared. Mining companies are digging deeper into the earth to find the resources that used to be relatively close to the surface. In some cases, the mining shafts extend

700 stories underground.[76] In 2008, the British-Australian mining company Rio Tinto established a "prototype mine" with an autonomous drill rig platform and autonomous haul trucks. In 2011, the company purchased 150 autonomous trucks, which will use AI to learn mine layouts, recognize potential obstacles, and detect other vehicles, preventing accidents. The AI uses GPS to calculate the most efficient paths in the mine, reducing travel and fuel consumption.[77]

- *Handling Components*: Kiva Systems has created a robot—the "Magic Shelf"—that can autonomously navigate a warehouse, finding necessary components or products and bringing them to the proper processing area. Companies like Zappos, Amazon, and Staples already use fleets of the AI workers—which resemble comically large, orange Roombas—to process online orders within giant warehouses.[78] Amazon considered the AI such an important cost-saving device that it bought Kiva in 2012 for $775 million in order to reduce labor expenses, even while expanding the number of its warehouse order centers.[79] Although Kiva's robots are primarily used in retail warehouses, the company believes that the AI can be used for components distribution in industrial settings.[80] Manufacturers are starting to take notice, as Foxconn—the manufacturing company most famous for producing iPhones en masse—has already halted hiring at its factories in anticipation that autonomous technology will be adopted in the next few years.[81]

- *General Factory Work*: Rethink Robotics is aiming specifically at the industrial sector with its general AI industrial worker, Baxter. Baxter is a humanlike AI that features a sonar system (to detect people and objects around it), five cameras for viewing its workspace, and two arms and hands for assembly line and other factory jobs. It is intended to be bolted to a work platform, and although it has wheels to be moved around the factory floor, it cannot move around on its own. It's not even equipped with Wi-Fi because most factories don't have it. Similarly, it can't recognize voice commands because factories are loud.[82] However, it can be trained—just grab its wrist and demonstrate what you want it to do. Baxter is such a good student that it can even learn from a drunk instructor.[83] Most importantly, it was designed to be a robotics platform, essentially meaning that it can support apps. Baxter's developers hope that it will become the robot iPhone, with programmers creating all sorts of apps for industrial functions (and many other functions, like bartending, possibly for Baxter's boozy trainers) that expand Baxter's responsibilities in factories.[84]

The AI described above is only a small sample of what is currently in development. AI industrial workers will expand beyond these developments, performing other industrial tasks and gradually reducing manufacturing's reliance on human workers even further.

How the Introduction of AI Industrial Workers (and Other Forms of AI) Affects Local Ordinances

AI industrial workers, AI farmworkers, and other forms of AI will enter towns and cities, interacting with municipal ordinances and zoning ordinances in ways that were not anticipated when municipalities wrote those laws. Ordinances that assumed large crowds of people will face far fewer people. Ordinances that sought to protect bucolic charm will ironically invite more machines and industry.

Compared with states and the federal government, though, towns and cities are better situated to address AI that makes decisions and navigates downtown like a person. The AI issues confronting municipalities are less abstract than those facing higher forms of government. Municipalities will not have to decide whether robots can be liable or what level of personhood to assign machines. Municipalities will have to treat AI like a person in easy-to-understand situations: where autonomous cars can park, which areas in town should be open to AI industrial workers, etc. Here are just a few examples of specific situations in which AI, in a variety of forms, challenges common municipal and zoning ordinances:

- Due to the increased use of autonomous cars, fewer cars park in downtown areas, as cars drop off their owners and head home. This results in a loss of municipal revenue. For example, Indianapolis collected $5.3 million from parking meters in 2012.[85] Boston collected $62.2 million from parking fines during fiscal year 2012.[86] Revenue from parking fines contributed $92.6 million to city revenue in Washington, D.C., during fiscal year 2011.[87] In Los Angeles, the city collects about $21 million a year from parking meters and $120 million a year in parking fines.[88] Although smaller towns collect smaller dollar amounts from parking fees, they are equally important to those communities.
- Advances in AI farmworkers permit small farms in rural and suburban areas to adopt autonomous technology, including tracks and fleets of robots to work in the fields. This defeats the purpose of zoning ordinance sections that encourage agricultural uses. Frequently,

towns want those sections to protect scenic rural vistas or create more attractive landscapes.[89]

- AI industrial workers Baxter and Kiva's Magic Shelf permit a boutique electronics manufacturer to operate a full programming and assembling plant in very limited space with little noise. The two owners—who are also the only employees—would like to house their company in a historic townhouse in the downtown of a midsize regional city that was rezoned for retail use when the city decided to encourage the growth of the downtown shopping scene. There will only be one delivery per week of components for assembly and programming and only one shipment per week of finished electronics. The nature of the production is fairly quiet, particularly for a downtown building. City leaders would like to attract similar small manufacturing companies and would also like the associated tax revenue. However, under the zoning ordinance, the commercial zone prohibits manufacturing. Even if the zone permitted manufacturing, the ordinance requires more parking spaces for manufacturing plants than are actually necessary for the two owners.

- AI industrial workers Baxter and Kiva's Magic Shelf also permit the owner of a fairly large manufacturer to operate an automotive parts plant in his suburban subdivision home. With the AI, he has been able to downsize his operation from 10 employees in a midrange industrial space to just him in his basement. He correctly noted that the "home office" provision of his town's zoning ordinance permits residents to operate businesses out of their homes if the business: has no customer or client visits, accepts no regular deliveries, maintains no specialized vehicles or exterior equipment, does not sell merchandise to the public on the premises, employs only people who live in the home, and does not display any exterior commercial sign.[90] Even though the manufacture of automotive parts involves loud grinding noises that disturb the neighbors, the use is permitted in the residential zone because it falls within the definition of "home business," which was originally intended for homeowners who operate small professional services businesses from their homes. The town does not have a noise ordinance.

- A college engineering student has reprogrammed his senior project—an AI milking machine for use on farms—into an AI sexual worker. (Popular fiction likes to imagine that AI sexual partners will be attractive and interactive like Jude Law in *A.I.* In reality, they will probably be similar to the International Space Station robotic arm that Howard

Wolowitz uses to masturbate in *The Big Bang Theory*.) He sells his design to an adult entertainment store nearby. The store is properly permitted under the zoning ordinance, which allows the "sale of sexually explicit services including . . . instruments, devices, or paraphernalia which are designed or used in connection with 'sexual conduct.'"[91] The store owner insists that the AI sexual worker falls under that language and is properly allowed in his shop. The City Council believes it is closer to prostitution, which is prohibited by the city ordinance.

How to Amend Ordinances to Incorporate AI

The scenarios described above are just examples, but they indicate the general problems AI will introduce to municipalities. Tax revenues associated with parking will decline. Towns that have carefully crafted their zoning ordinances to protect their identities will find that AI can limit the effectiveness of their ordinances to do that. AI will make it easier for manufacturers to move into nonindustrial areas without disturbing neighbors, but many existing ordinances will make it difficult for manufacturers to do that. Ironically, it will also be easier for nonresidential uses to occur in residential areas, bothering neighbors. People will be able to use AI for adult entertainment that would be illegal if that entertainment were with another person. The strategy that municipalities should embrace is to include AI behavior in the human behavior governed by local ordinances, giving AI rights and obligations in certain situations.

For example, parking meters and fines address the location and duration of people's parking. Cities and towns should revise their ordinances to address the behavior of autonomous cars. To compensate for lost parking revenue, municipalities may have to consider a toll similar to the congestion toll Mayor Michael Bloomberg suggested in 2007 for New York City. The plan called for tolls into the city to vary depending on the time of day, similar to what London has used since 2003. When congestion on the roads leading into the city is high, tolls would be higher. During weekends and nonwork hours, the tolls would be reduced.[92] In municipalities that notice a dramatic increase in autonomous cars and a corresponding decrease in parking revenue, a similar system can be implemented, assessing a toll for vehicles entering the downtown or commercial area during certain times of day, probably the times that are currently subject to parking meter requirements. Towns and cities can collect the tolls through an EZ-Pass or similar device.

Municipalities will have to calculate the toll carefully. They cannot replace parking meter revenue with the toll one-for-one because parking revenue includes the fines cities and towns collect from cars that have overstayed or not paid meters. Municipalities must calculate lost revenue from fines as well as lost parking meter revenue. Additionally, the point is not to charge residents who live in the downtown or commercial areas, so ideally the cars of residents will receive a waiver.

Similarly, ordinances now dictate how people use the land in order to protect scenic or rural areas, but municipalities should consider monitoring the development of AI agricultural workers. At a certain point, they may want to consider dictating how AI uses the land to protect scenic and rural areas. Are tracks crisscrossing farmland acceptable? Can robots operate farms 24 hours a day? Should municipalities limit the size of AI agricultural workers? These are elements of AI farmworkers that can seriously impact the rural character of a bucolic town or a town's scenic views.

Additionally, many zoning ordinances do not address the potential for manufacturing and industrial uses in retail spaces. AI will make manufacturing palatable in areas where it would not be if 50 people were working machines because only a few people, if any, will be required to perform the same work. "More people" equals greater use of property by those people, increasing the potential to bother neighbors. That is one of the reasons ordinances try to limit different uses—commercial, manufacturing, etc.—adjacent to one another.

As AI industrial workers enter more factories, municipalities should consider revising the uses allowed in commercial, retail, and downtown districts to permit some manufacturing, depending on what the AI does. Quiet AI that does not create noise, odors, or other negative externalities should be permitted, regardless of what the people occupying the property use it for. This will give downtowns and other commercial areas greater flexibility to develop as economic engines.

Rethinking downtown in this way is similar to what many former mill towns in New England did in the late 20th century, after local manufacturers closed their factories. These towns frequently started as company-owned, planned communities in the 19th century. The mills were the center of the community economically, geographically, and architecturally. When the mills closed, the towns that had grown around them needed to reinvent themselves.[93] Many revised their zoning ordinances to permit commercial and residential uses in the former industrial buildings to encourage economic growth in the buildings that were essentially the center of town.[94]

Manchester, New Hampshire, for example, was founded by the Amoskeag Manufacturing Company in the 1830s when it started a textile factory along the shore of the Merrimack River.[95] By the early 20th century, the sprawling Amoskeag facility in Manchester was the largest manufacturer in the world, and Manchester had 75,000 residents.[96] But pressure from factories in southern states and other countries compounded by the Depression forced Amoskeag to close on Christmas Eve 1935.[97] Although civic leaders had some success promoting the mill buildings to other manufacturers for a while,[98] that area of Manchester became a greater economic force in the late 20th century when the city rezoned the mill buildings for mixed use, permitting offices, schools, stores, and restaurants in Amoskeag's former space.[99] Manchester's experience was not unique.[100] Ironically, towns and cities should now consider rezoning in reverse, from commercial to manufacturing, in order to ensure that their downtown areas are open to AI and remain or become drivers of economic growth.

On the other hand, municipal leaders want to protect homes from inappropriate economic activity, like obnoxious industrial uses in someone's basement. As AI becomes more common, towns should revisit their ordinances to make sure the sections governing home occupations address what AI does on the property in addition to what the homeowner does. Homeowners may be authorized to have a business in their homes, but only to the extent that it does not interfere with their neighbors. In fairness to towns and cities, most current zoning ordinances have language that tries to make the character of the neighborhood a consideration or that requires any home occupation to be unobtrusive. But a review of these sections in light of AI development will help towns regulate AI behavior in subdivisions and residential neighborhoods.

In some cases, towns and cities may decide that AI behavior should be governed in exactly the same way that human behavior is governed. AI prostitutes present a likely candidate. I suspect that a lot of city leaders will not be comfortable with a city ordinance that bans humans selling sex but permits AI to sell the same thing (even if no one will argue that prostitution is degrading for the AI) because of increased crime and other secondary effects associated with adult entertainment businesses.[101] Although the Supreme Court has ruled that municipalities may not ban adult entertainment establishments entirely due to freedom of speech protection afforded by the First Amendment,[102] municipalities will likely have more success revising their municipal ordinances to ban AI from performing as a sex worker because of the secondary effects associated with traditional prostitution and the fact that if a human prostitute does not have a First Amendment interest in prostitution, the AI certainly does not. However, municipalities should be aware that depending on the circumstances, obtaining convictions for solicitation can be more difficult than successfully enforcing the zoning ordinances that typically regulate the use of property for adult entertainment businesses, as it is more difficult to conceal a store in a building than an individual's activity.[103]

AI before the Board of Selectmen

Addressing the potential problems identified in this chapter does not represent an impossible challenge. In many cases, the problems will not even be especially difficult for towns and cities. In some ways, municipalities are better suited to address the legal issues AI will create for them than are states and the federal government. Municipalities govern discrete and local issues: public education, streets, town parks, sewers, etc. Cities will not have to determine overarching rules for liability and user responsibility associated with AI like states; towns will not have to address how AI will change intellectual property and international laws like the federal government. This role is consistent with the historical development of municipal and zoning ordinances, and those tools should be sufficient to address AI, too. Like state and national leaders, town and city leaders will need to remember that their ordinances must treat AI more like human beings, revising them to address the behavior of AI.

Although municipalities will generally have more manageable decisions to make about AI than either state or federal governments, one area where they will all overlap is AI in law enforcement. Municipal police departments will have to be aware of how the Fourth Amendment limits their use of AI surveillance drones, but the states, the federal government, and the courts systems will provide input and guidance. I'll explore this in the next chapter.

CHAPTER 6

AI and the Fourth Amendment

Perhaps more than any other area of the law, technological changes have shaped the Fourth Amendment of the United States Constitution, which protects people and their "houses, papers, and effects" from "unreasonable searches and seizures" without a warrant that has been issued based on "probable cause." When the Founding Fathers wrote and ratified the Fourth Amendment, there were no phones, computers, Internet, or cars. There was no way to measure heat from a distance, to record sound, or to take pictures from the air. As a result, police, courts, and government officials have had to use broad 18th-century language to address increasingly complex personal spaces and increasingly sophisticated means to search them.

Consider just a few recent court cases that have examined searches that would not have been possible during the American Revolution:

- *Kyllo v. United States*—The U.S. Supreme Court considered whether police needed a warrant to examine a house from an off-site location using a thermal imaging device that could detect heat associated with lights for growing marijuana.[1]
- *Commonwealth v. Phifer*—The Massachusetts Supreme Judicial Court looked into the search of a suspect's cell phone following his arrest.[2]
- *United States v. Jones*—The U.S. Supreme Court considered whether police needed a warrant to attach a GPS tracker to a suspect's car and track his movements.[3]

James Madison had not considered the protection of home pot farms when he marshaled political support for the Fourth Amendment, but then again, he did not think of preventing home pot farms either.

There are two technological movements that have strained the Fourth Amendment. The first is the development of new devices that police can use to collect information about crimes and criminals, such as thermal imaging devices and GPS trackers. The second is the development of new programs and machines that expand the spaces people use to store information, such as cell phones, as well as the spaces that gather information whether we want them to or not. Although AI will introduce more devices that will collect data about us—How does he use his AI nanny? Where does she go in her autonomous car?[4]—this chapter will focus on the first technological movement, the development of new police technologies.

Police are already experimenting with surveillance drones that can watch, follow, and search for people in ways that were once impossible, or at least prohibitively expensive. Soon we'll see AI surveillance drones that can monitor us from the air and on the street without a human directly operating it. Although not Gort of the Interstellar Police from *The Day the Earth Stood Still*, these drones represent very sophisticated weak AI that will test the Fourth Amendment.

In order to fully appreciate that challenge, it is helpful to review the Fourth Amendment and how court decisions, police actions, and technology have shaped it in the last 200+ years. It will also be helpful to explore historical, current, and expected use of robots and AI by police. The next two sections focus on those discussions. The final section of this chapter looks at AI that law enforcement will use and considers it through our current understanding of the Fourth Amendment.

Brief History of the Fourth Amendment

Although the Fourth Amendment dates to 1791, its origins predate the American Revolution, but the U.S. Supreme Court waited nearly a century before interpreting it. Once Fourth Amendment questions appeared on the Court's docket, new technologies forced the justices to frequently reevaluate the protections granted by the amendment.

The Founding Fathers' Perspective on Freedom from Unreasonable Search and Seizure

Fresh in the minds of the Founding Fathers was the idea of general warrants and abusive searches. They were aware that King George III permitted his agents to arrest anyone and search houses in order to locate libelous authors and their papers.[5] The English courts rejected this practice, with the opinion written by Lord Camden stating that searching for evidence in a man's home without a specific warrant is cruel and unjust; both the guilty and innocent will be punished.[6] During the French and Indian War, many of the Revolutionary leaders experienced firsthand the broad authority granted to English revenue and customs officers to search suspected places for smuggled goods.[7] Parliament antagonized the colonies when it passed the Townshend Act of 1767, which reauthorized customs officials to search any places suspected of hiding smuggled goods and led to unrest up until the eve of the Revolutionary War.[8] Many from that generation believed these searches were what first pushed the colonies to declare their independence. As John Adams noted: "Then and there was the first scene of the first act of opposition to the arbitrary claims of Great Britain. Then and there the child Independence was born."[9]

With this recent history in the minds of Adams's contemporaries, states demanded an amendment to the new American Constitution that would protect this right. Massachusetts, New York, and Virginia were particularly vocal in this demand, leading to the inclusion of the Fourth Amendment in the Bill of Rights in 1791.[10] They were primarily concerned with ensuring that police and government searches and arrests would be reasonable and limited to instances when a specific warrant was obtained.[11]

Development of the Fourth Amendment in the Courts

As the Fourth Amendment is constantly in front of the courts today, it might seem strange that the Supreme Court did not issue a ruling on it until *Boyd v. United States* in 1886. This is due to the fact that during the 18th and 19th centuries, courts applied the Bill of Rights only to the federal government, not to the states. Additionally, states enacted most of the criminal laws, and most of the criminal prosecutions occurred in state courts, so there were few opportunities to apply the Fourth Amendment to federal police powers.[12] However, every state had a constitutional or statutory provision that applied the rights in the Fourth Amendment to its law enforcement actions.[13]

In *Adams v. New York* from 1904, the Supreme Court confirmed that the Fourth Amendment, in and of itself, did not apply to the states, but expressed some doubt as to the effect of the Fourteenth Amendment, which may make the Fourth Amendment applicable to the states.[14]

In 1961, the Court ruled in *Mapp v. Ohio* that under the Due Process Clause in the Fourteenth Amendment, which states that no "State [shall] deprive any person of life, liberty, or property, without due process of law," the Fourth Amendment applies to the states because "'security of one's privacy against arbitrary intrusion by the police' is 'implicit in the concept of ordered liberty' and, as such, enforceable against the States through the Due Process Clause."[15] This meant that the Supreme Court could review the actions of state law enforcement and determine what constitutes unreasonable searches and seizures. In *Mapp*, the Court used that authority to rule that states cannot use evidence in court if it was obtained through a search or seizure that did not comply with the Fourth Amendment.[16]

In 1967, Justice John Marshall Harlan's concurring opinion in *Katz v. United States.* established a two-part test that continues to be the standard courts use to determine if law enforcement officers have actually conducted a search and seizure under the Fourth Amendment.[17] Harlan wrote that "there is a twofold requirement, first that a person have exhibited an actual (subjective) expectation of privacy and, second, that the expectation be one that society is prepared to recognize as 'reasonable.'"[18] He continued, noting: "A man's home is, for most purposes, a place where he expects privacy, but objects, activities, or statements that he exposes to the 'plain view' of outsiders are not 'protected,' because no intention to keep them to himself has been exhibited. On the other hand, conversations in the open would not be protected against being overheard, for the expectation of privacy under the circumstances would be unreasonable."[19] In other words, if the police search your bag, you must prove that (1) you thought your bag was private, and (2) that was a reasonable idea in order to prove that the police engaged in a Fourth Amendment search. So if the police search the contents of your bag after you have dumped them in the middle of the sidewalk, that is not a search for the purposes of the Fourth Amendment because you did not act like the contents of the bag were private. On the other hand, if you placed the bag in your car and locked the doors, a police search of the bag would qualify as a search under the Fourth Amendment because you acted like it was private and that was a reasonable assumption.

Once you have established that the police actually conducted a search and seizure under the Fourth Amendment, then you can attempt to prove that the police violated your Fourth Amendment rights. In a criminal case, this typically requires showing that there was no warrant because, except in well-defined circumstances, a search and seizure is considered unreasonable if there is no warrant issued upon probable cause.

However, exceptions exist when "special needs, beyond the normal need for law enforcement, make the warrant and probable cause requirement impracticable."[20]

A few examples from actual Supreme Court cases might be useful. In *Skinner v. Railway Labor Executives Association*, the Court ruled that random urinalysis drug tests required by the Federal Railroad Administration (FRA) qualified as a Fourth Amendment search.[21] Justice Anthony Kennedy noted that urine may reveal sensitive medical information and that collecting the sample involves another person watching or listening to another pee, which "itself implicates privacy interests."[22] However, because the FRA used the drug tests to prevent train accidents caused by drugs or alcohol and not to prosecute employees, the tests were reasonable under the Fourth Amendment and did not violate the employees' constitutional rights, even though there was no warrant or probable cause.[23]

Conversely, in *Smith v. Maryland*, the Court ruled that there was no Fourth Amendment search involved when the police installed a pen register onto a suspect's telephone without a warrant.[24] A pen register records all of the numbers dialed from a particular phone, but not the contents of the telephone conversations.[25] Because callers provide the numbers they call to the phone company, they have no reasonable expectation of privacy, regardless of whether they think of those numbers as private. As a result, the pen register was not a search and was not protected by the Fourth Amendment.[26]

How Courts Have Applied the Fourth Amendment to New Technologies

This is the legal background with which police and courts have had to comply as they have integrated new devices into their practices and as the population at large has adopted new technologies. Some of these developments suggest how courts and police departments will have to adjust to accommodate AI in law enforcement. Let's start by looking at the cases that opened this chapter: *Kyllo v. United States*, *Commonwealth v. Phifer*, and *United States v. Jones*.

In *Kyllo*, agents from the U.S. Department of the Interior suspected that Danny Kyllo was growing marijuana in his home. Knowing that the lamps required to grow pot indoors create a great deal of heat, they used a thermal imager, which was set up in the backseat of their car, to scan Kyllo's home for a few minutes. The scan showed that the area around Kyllo's garage was warmer than the rest of his house and substantially warmer than the neighboring homes in his triplex. The agents concluded

Kyllo was growing pot in his home. Relying on the thermal scan and tips from informants, a judge issued a warrant for his arrest. Kyllo was indicted on one count of manufacturing marijuana, but tried to suppress the thermal imaging evidence.[27]

The Supreme Court stated that with few exceptions, "[T]he question whether a warrantless search of a home is reasonable and hence constitutional must be answered no."[28] In his majority opinion, Justice Antonin Scalia pointed to three cases that addressed the use of airplanes to conduct surveillance:[29]

- *California v. Ciraolo*, where the Court ruled that aerial surveillance of a fenced-in backyard from an altitude of 1,000 feet in public airways did not constitute a search because there is no reasonable expectation of privacy in an area that can be readily observed by commercial flights;[30]
- *Florida v. Riley*, where the Court ruled that aerial observation of the partially exposed interior of a residential greenhouse from 400 feet above in a helicopter did not constitute a search because helicopters are permitted to fly at that height by the FAA;[31] and
- *Dow Chemical Co. v. United States*, a case in which the Supreme Court ruled that enhanced aerial photography of an industrial complex did not require a warrant.[32]

Using these cases, Scalia explained that naked-eye observation of homes from the air, as in *Ciraolo* and *Riley*, is acceptable, but that using technological enhancement of ordinary perception to examine a private home where the resident has clearly expressed an expectation of privacy is "too much."[33] With this background, the Court determined that when "the Government uses a device that is not in general public use to explore details of the home that would previously have been unknowable without physical intrusion, the surveillance is a 'search' and is presumptively unreasonable without a warrant."[34]

The *Phifer* case comes from the Massachusetts Supreme Judicial Court, the state's top court, and addresses an issue that the Supreme Court has not yet considered: whether the recent call list contained within a cell phone may be searched as part of a lawful arrest.[35] Demetrius Phifer was charged with distribution and a drug violation near a school or park. He moved to suppress evidence obtained from a warrantless search of his cellular telephone, which the arresting officers obtained after arresting him. He attempted to argue that the search was in violation of the Fourth Amendment because the officers did not have a warrant, probable cause, or his consent.[36] The SJC relied in part on the "search-incident-to-arrest" exception to the Fourth Amendment—which permits searching a suspect as part of an arrest for weapons, instruments of escape, and evidence of crime—to rule that the search was constitutional because the cell phone was properly taken during a permissible search as part of his lawful arrest and the search was limited to the list of recent calls.[37] However, the court limited its ruling, so as not to unreasonably expand the government's constitutional search powers into a relatively new technology, saying that the decision does not control "in relation to a different type of intrusion into a more complex cellular telephone or other information storage device."[38]

The *Jones* decision considered police use of GPS without a proper warrant. The Federal Bureau of Investigation and District of Columbia police suspected Antoine Jones of trafficking in narcotics and wanted to track his movements. Although the U.S. District Court of the District of Columbia issued a warrant permitting law enforcement to attach a GPS to Jones's car, the warrant stated that the GPS must be installed in the District of Columbia and within 10 days. Law enforcement attached the GPS in Maryland on the 11th day.[39]

The GPS device was installed on Jones's car for 28 days. During that time the government tracked the vehicle's movement's and was able to determine its location within 50–100 feet. The GPS communicated the car's location by cell phone to a government computer. By the end of the 4 weeks, it had produced more than 2,000 pages of data and evidence against Jones. Jones was ultimately arrested and convicted for conspiracy to distribute and possess cocaine. The prosecutors relied, in part, on the evidence gathered by the GPS.[40]

On appeal, Jones's conviction was overturned. The majority opinion for the Supreme Court stated that "the Government trespassorily inserted the information-gathering device" and that such a trespass violates the Fourth Amendment's requirement that the people be "secure in their persons, houses, papers, and effects."[41]

However, the concurring opinions of Justices Sonia Sotomayor and Samuel Alito address the technological challenge of GPS more directly. Sotomayor describes the potential dangers involved in unchecked government use of GPS devices: "GPS monitoring generates a precise, comprehensive record of a person's public movements that reflects a wealth of detail about her familial, political, professional, religious, and sexual associations. [Citation omitted.] Government can store such records and efficiently mine them for information years into the future. [Citation omitted.] And because GPS monitoring is cheap in comparison to conventional surveillance techniques and, by design, proceeds surreptitiously, it evades the ordinary checks that constrain abusive law enforcement practices: 'limited police resources and community hostility.' "[42] She considers what we believe to be private, wondering about: "the existence of a reasonable societal expectation of privacy in the sum of one's public movements. I would ask whether people reasonably expect that their movements will be recorded and aggregated in a manner that enables the Government to ascertain, more or less at will, their political and religious beliefs, sexual habits, and so on."[43]

Justice Alito compares new information-gathering technologies with previous investigative techniques: "In the pre-computer age, the greatest protections of privacy were neither constitutional nor statutory, but practical. Traditional surveillance for any extended period of time was difficult and costly and therefore rarely undertaken. The surveillance at issue in this case—constant monitoring of the location of a vehicle for four weeks— would have required a large team of agents, multiple vehicles, and perhaps aerial assistance. Only an investigation of unusual importance could have justified such an expenditure of law enforcement resources. Devices like the [GPS] used in the present case, however, make long-term monitoring relatively easy and cheap."[44] He worries that such technology "in investigations of most offenses impinges on expectations of privacy" because "society's expectation has been that law enforcement agents and others would not— and indeed, in the main, simply could not—secretly monitor and catalogue every single movement of an individual's car for a very long period."[45]

The lesson to take from these cases as we turn to focus on AI is that, in general, new technologies that permit enhanced surveillance above and beyond observations made with the naked eye will be treated skeptically. This is especially true where technology permits prolonged surveillance because even though people participate in the public sphere regularly— going to work, traveling to friends, running errands, etc.—they believe that the sum total of those public activities is private. And even when courts are willing to permit law enforcement agents to search using newer technologies, such authorization will be very limited initially due to concerns related to abusive searches in technologies not entirely understood.

With this approach to new technologies in mind, let's look at how police are incorporating robots and AI into their efforts to make our communities safer.

Police Use of Robots and AI

Although the use of robots in police work is relatively new, the serious consideration of robots in law enforcement goes back to at least the

1950s, when scholars began to speculate about robots that will "take over the simple, semi-mechanical tasks, but slowly, then swiftly," so that "millions of clerks will be freed for other work by the astonishing adaptability of this magnificent autocrat."[46] In 1972, British Lieutenant Colonel Peter Miller invented a remote-controlled robot for bomb disposal by the U.K. Army in response to improvised IRA bombs in Northern Ireland.[47] By the early 1990s, police departments had begun to incorporate these large, remote-controlled robots (on the order of 1,000 pounds) for bomb removal.[48] The robots enter situations that could be lethal to human police officers should the explosive ignite. However, they do not always operate exactly as intended. In 1993, a large bomb-removal robot nicknamed "Snoopy" malfunctioned, "spinning around, just going wild" while attempting to defuse an explosives situation. Fortunately, Snoopy had not yet grabbed the pipe bomb it was supposed to obtain, and human bomb squad officers removed the device by hand and detonated it in a remote area.[49]

Bomb removal continues to be the most common use for a police robot.[50] However, more recently, the law enforcement community has expanded its use of robots into investigation,[51] street monitoring,[52] and hostage negotiations.[53] In April 2013, robots assisted the Boston Police Department while officers searched for the Boston Marathon bombers.[54] Some police departments are looking into using armed robots with their SWAT teams.[55] At this point, all of the robots in police use are controlled by human beings.[56] And they continue to have some bugs, some of which can be pretty ridiculous, sort of like the inability of ED-209 from *Robocop* to navigate stairs (a serious shortcoming for a security drone).[57] But at least ED-209 was waterproof. In 2007, the police department in Perm, Russia, deployed a robot called R Bot 001 to monitor street crime. It featured a button for citizens to use to contact the police station and the ability to deliver simple orders like telling drunk Russian pedestrians to go home and sleep it off (to use an example that is both real and stereotypical). Unfortunately, R Bot 001's designers overlooked some basic outdoor conditions. During the first rainstorm of its patrol, the water shorted the electrical system.[58]

But the robots also continue to have legitimate benefits. In March 2013, an inebriated man armed and barricaded himself in his Ohio home. After a human negotiator struck out, the police sent in two investigative robots. The man shot—but did not destroy—one of those robots, after which the police entered the home and arrested him.[59] Police departments are eager to take advantage of surveillance drone technology as well.[60] In 2006, the Los Angeles Sheriff's department began flying small drones to track suspects, but the Federal Aviation Administration (FAA) forced the department to terminate the practice,[61] annoying several senior officers, who alleged that the FAA was inconsistent in how it permitted government agencies to use drones.[62]

The FAA is charged by Congress with regulating U.S. airspace,[63] so police departments, sheriffs' offices, and other law enforcement agencies must comply with its rules governing drone use. The first unmanned aircraft was authorized by the FAA in 1990.[64] Currently, the FAA distinguishes among unmanned aircraft that is used by public entities, by private entities, and as model planes.[65] With regard to public and private entities, the FAA has developed an authorization system, in which public entities (including public law enforcement agencies) may apply for a Certificate of Waiver or Authorization (COA) and private entities may apply for a Special Airworthiness Certificate (SAC). COAs and SACs are issued for as long as a year.[66] Public entities are required to state the purpose of the desired aerial drones, the type of drones, and the geographic area where they will be used.[67] Private entities typically only receive a SAC if they are developers or manufacturers looking to use the aerial drones for research, development, testing, crew training, and market surveying;[68] other commercial uses for aerial drones are not authorized by the FAA.[69] Model airplane enthusiasts are cool—they don't need FAA approval,[70] but are advised to fly their planes within 400 feet of the ground and no closer than three miles from airports without notifying the airport operator to avoid aircraft in flight.[71]

The FAA claims limited legal right to release information regarding issued, pending, or expired COAs and SACs, although in February 2013, it issued a press release stating that it had given out 1,014 COAs between 2009 and 2012 and that 327 were still active as of February 15, 2013.[72] There have been numerous Freedom of Information Act requests to the FAA asking for the entities that have applied for authorization and those that have received authorization, providing more detailed information about entities authorized to operate aerial drones in the United States.[73]

According to information released pursuant to those requests, between November 2006 and June 30, 2011, there were 13 local law enforcement agencies that were authorized to operate aerial drones.[74] Through October 2012, the FAA has received applications from 17 local law enforcement agencies to use aerial drones.[75] It is not clear exactly how many aerial drones are currently authorized in the United States by the FAA. Relying on a Government Accountability Office report,[76] the *Los Angeles Times* reported in February 2013 that the FAA has issued 1,428 permits to domestic drone operators since 2007.[77] It is unclear if these numbers are consistent with the numbers the FAA released that same month, but they are not entirely consistent with information the FAA has provided to Congress and released pursuant to Freedom of Information Act requests.[78]

There might be more confusion coming. In May 2012, the FAA provided a more streamlined application process for police departments and other governmental agencies who would like authorization to use aerial drones, although there are still restrictions regarding the size of the drone and the altitude where they can operate.[79] When President Obama signed the FAA Modernization and Reform Act of 2012,[80] it required the FAA to develop regulations to permit commercial use drones beyond research and testing by 2015.[81] Advocacy groups worried that the proliferation of unmanned aerial drones—the FAA predicts that there could be 30,000 in American public airways by 2020—will wreak havoc on personal privacy, particularly if the FAA authorizes many more law enforcement agencies to use drones but does not release information about that use. Jennifer Lynch of the Electronic Frontier Foundation has said, "We need a list so we can ask [each agency], 'What are your policies on drone use? How do you protect privacy? How do you ensure compliance with the Fourth Amendment?' "[82] Although there is no reason to believe the FAA is unconcerned with these privacy issues, the agency's primary concern has to be more practical: how do you keep so many flying objects from crashing into each other and plummeting into our houses?[83] This concern is already real, as pilots of commercial airlines have begun reporting near collisions with drones.[84]

Advocacy groups are not the only entities that are worried about privacy. States and municipalities have started working on legislation and ordinances to ban drones. In 2013, Florida enacted a law that bans law enforcement from using aerial drones except in cases involving terrorism, imminent dangers like missing or endangered people, and search warrants.[85] The sponsor rebuffed efforts by Florida police to expand the exceptions to include crowd control.[86] At least thirteen states have begun considering similar bills that would force police departments to obtain search warrants before using aerial drones for surveillance.[87]

In February 2013, Virginia's legislature went even further, passing a two-year moratorium on the use of aerial drones by state and local law enforcement agencies,[88] just one day after the City Council in Charlottesville, Virginia, passed a similar resolution banning city use of drones for two years.[89] Shortly thereafter, Seattle Mayor Mike McGunn ordered the Seattle Police Department to abandon its plan to use drones after residents protested.[90] In 2012 and 2013, Indiana considered a bill that would prohibit *all* aerial drones, public and private, within its borders, although that bill did not become law.[91]

Introduction of AI Drones

Bear in mind that the surveillance drones causing consternation among local and state governments and advocacy groups are all human-operated. None of them are self-flying, none use AI. But that technology is coming.

DARPA—the federal agency that has sponsored or performed pioneering AI work with Siri, autonomous cars, and robotic nursing assistants—is working on perfecting numerous elements of autonomous drone flight, including midair refueling[92] and payload placement.[93] (Autonomous drones that can fly over hostile enemy territory performing attacks or targeted killings are discussed in the next chapter, but in the same way that bomb-removal robots trickled down from the military to local law enforcement, the current movement of drone technology suggests that autonomous military drones will lead to autonomous law enforcement drones.) The federal government has already begun limited use of AI drones. For example, in 2008, the National Oceanic and Atmospheric Administration experimented with a mostly autonomous aerial drone (it only used human operators to land and take off) to scout for fishing nets adrift at sea, which kill thousands of birds and marine mammals each year. The drone was equipped with video sensors that could detect anomalies in the water and with GPS sensors that mark the location of such anomalies for later recovery by boats.[94]

Autonomous drones may be ready for us to purchase by 2014, when AI called the "MeCam" is tentatively scheduled to hit the market.[95] Always Innovating, the developer of the MeCam, notes that the AI "launches from the palm of a hand and hovers instantly … The MeCam doesn't need any remote control: the user can control the device with voice commands or use the follow-me feature."[96] The "follow-me feature" permits the MeCam to follow the owner/operator autonomously. The MeCam is equipped with a camera, and users will be able to launch the device easily, give it instructions about where to hover and when to film, and record live video directly to their smart phones, computers, or tablets.[97]

If Always Innovating hopes the MeCam will appear on the market before the FAA's new regulations in 2015, it will have to address the FAA's current prohibition of most private, commercial drones. But the company's confusion is understandable. In a 2009 report, the FAA indicated that only 30 percent of its regulations are clearly applicable to aerial drones; the remaining 70 percent either are not relevant or are unclear as to whether or not they apply to aerial drones.[98] So although the FAA has stated publicly that it does not authorize the private use of drones for anything other than research and development, companies can be forgiven if they disagree with such a hard and fast interpretation of the regulations. U.S. law actually says very little about unmanned aircraft. In fact, as of early 2013, no variation of unmanned aircraft, unmanned aerial vehicle, or aerial drone (which essentially all mean the same thing) has been codified by a law passed by Congress or a regulation promulgated by the FAA.[99] The regulations that the FAA points to for support of its policy toward private aerial drones are actually intended to govern *manned* experimental aircraft.[100] They do not specifically address unmanned drones.

And they say even less about autonomous aerial drones. Although the FAA vaguely refers to unmanned aircraft that are flown "autonomously through use of an on-board computer" in a FAQ addressing unmanned aircraft generally, the FAA has not specifically addressed autonomous aerial drones as compared to human-operated drones.[101] As the 2012 FAA Modernization and Reform Act did not specifically address autonomous aerial drones,[102] it is not clear whether the new regulations in 2015 will address this issue as well or whether they will leave autonomous unmanned aircraft ambiguously regulated.

Such ambiguity means that as of 2015, law enforcement agencies could have more leeway to use drones like the MeCam. How will courts view AI surveillance drones like MeCam under the Fourth Amendment? Will warrants be required any time the police want to use AI drones, or will there be general exemptions for particular scenarios, like missing persons?

AI Surveillance Drones under the Fourth Amendment

The question of constitutionality cannot be definitively answered by Congress. Rather, we depend on the courts to review government actions to ensure that they do not violate the Constitution. Courts will have the same role when AI surveillance drones are considered under the Fourth Amendment.

How Past Court Cases Will Inform Future Cases Reviewing AI

Although there is no way to predict how courts will rule on future cases, we can learn a lot about how courts might analyze AI surveillance drones by looking at recent cases that apply the Fourth Amendment to other forms of new technology police have used. In particular, the cases we discussed above—*Kyllo*, *Phifer*, *Jones*, *Ciraolo*, *Riley*, and *Dow Chemical*—suggest how courts will look at AI surveillance drones. By applying the general principles from those cases, we can make educated guesses about how the Fourth Amendment will limit law enforcement's use of these drones. Although I mentioned those principles in passing above, a more thorough review is useful before discussing AI and the Fourth Amendment.

- *Courts will continue to look at whether there was a subjective expectation of privacy and whether that expectation was reasonable when determining the applicability of the Fourth Amendment to AI activity on behalf of government agencies.* The *Katz* test that all of the cases above relied on—either explicitly or implicitly—will continue to govern cases in which a defendant claims that the government violated his or her Fourth Amendment rights. That is, courts will first ask if the defendant expected privacy in that particular situation; assuming the defendant has that expectation, courts will next ask if that expectation was reasonable. So, if a warrantless AI drone sneaks into a defendant's home and hides in his closet to record conversations between him and his wife, that will almost certainly violate the defendant's Fourth Amendment right to be free from unreasonable search and seizure. But if a warrantless AI drone records a conversation between the defendant and a coconspirator while walking in a public park, it is unlikely that a court will agree with the defendant's Fourth Amendment claim.
- *Courts will tightly limit their initial rulings governing AI and the Fourth Amendment.* In *Phifer*, the Massachusetts Supreme Judicial Court emphasized that its decision, which approved the search of the recent call list contained within an arrested suspect's cell phone, would not apply to other scenarios, other phones, other devices, or other information. The court's concern was that its decision would govern technological advances that it did not foresee in that decision and that the decision would govern such advances badly. Similarly, any court that reviews searches performed by AI drones will be careful not to hinder future AI developments with an overreaching decision based on early AI technology.
- *Unenhanced human observation from public areas does not require a search warrant.* *Ciraolo* and *Riley* suggest that unenhanced human observation does not constitute a search under the Fourth Amendment. In both cases, the Supreme Court looked at human observers who found incriminating evidence while above the earth in an aircraft (a plane and a helicopter). Because the human observers were in public airways and did not use technology to enhance their ground views, the Court ruled that there was no "search" as

that word is used in the Fourth Amendment and thus the evidence was admissible in court because no warrant was needed.

- *Observations from public spaces that are only possible through technological enhancement may require a search warrant* if *the place observed is a home.* In *Kyllo,* federal agents obtained an arrest warrant and conviction of a suspect in part by relying on observations of a home, from a public street, made possible with a thermal imager. In *Dow Chemical,* federal agents relied on observations of an industrial site, from the public airway, made possible with aerial photography. The Supreme Court ruled that the use of the thermal imager to gather information about a home constituted a search and that the evidence obtained from that search should be excluded because there was no warrant for that search. The Court ruled that the use of aerial photography to gather information about an industrial site was not a search prohibited by the Fourth Amendment. The difference appears to have been that one search targeted a home—which is specifically listed among the protected areas of the Fourth Amendment, "persons, houses, papers, and effects"—and one targeted an industrial property, which does not receive the same level of constitutional protection.

Although the special status of a residence under the Constitution was certainly a factor that distinguished these cases from one another, the Court may point to another factor in future cases looking at how the Fourth Amendment limits government use of AI surveillance drones:

- *Observations from public spaces that are only possible through technological enhancement may require a search warrant if the technology is not in general public use.* The Court in *Dow Chemical* noted that aerial photography is in widespread use and anyone flying in the public airway above the property could have taken those photos. In contrast, the thermal imager in *Kyllo* was not generally available to consumers, which the Court pointed out.
  It is an open question which factor would be more important to courts reviewing police use of AI surveillance drones: the search of a home or the use of commonly available technology.
- *Cumulative public actions can be private.* Although the Supreme Court's majority opinion in *Jones* somewhat inexplicably focused on the police trespassing on the suspect's car to place the GPS device, both concurring opinions addressed prolonged surveillance, which will likely make them more influential in future cases than Justice Scalia's odd ode to 18th-century common law. Justices Sotomayor

and Alito note that prolonged surveillance potentially infringes on the privacy interests protected by the Fourth Amendment, even when the prolonged surveillance occurs entirely in public, nonprotected areas. Such surveillance can indicate a person's religion, political beliefs, sexual orientation, etc., which are certainly traditional topics within the "persons" protected by the Fourth Amendment. Additionally, whereas prolonged surveillance has historically been so expensive that only the most important investigations warranted it, the advent of AI drones suggests that cheap, near-constant surveillance in any investigation may soon be possible.

Again, there's no way to know for certain how courts will apply the Fourth Amendment to any future case, let alone one that involves groundbreaking AI. But there is no reason to believe that these principles will not be among the most important that courts use when considering that AI.

Court Consideration, Unique to AI Surveillance Drones

In addition to the principles above, I would also suggest that courts should at least consider whether the AI's actions would have violated the Fourth Amendment if the AI were not a machine but a human police officer. One of the key benefits of AI surveillance drones, like the MeCam, is that a government officer can provide a simple instruction—"Follow me ... Follow him ... Fly to this location and wait to follow anyone who leaves"—and does not have to manually operate the drone. In many ways, this delegation of responsibility to the AI mimics the delegation of responsibility to another officer. Both the AI and other officer will exercise their own judgment within the confines of their orders. Both will be supervised by the delegating officer. Both can return with evidence leading to an arrest and conviction. There is a certain logic in courts reviewing the AI's actions as they would a human officer's actions in the same circumstances. This is also consistent with the concurring opinions in *Jones*, which noted the importance of taking the attributes of new technological developments into consideration when reviewing them for Fourth Amendment compliance.[103]

Potential Cases Involving Police Use of Pending AI

With these general principles and considerations in mind, let's consider a few police investigations that would be likely to utilize the MeCam or a similar device and review what a court is likely to say about the MeCam's compliance with the Fourth Amendment. For convenience, these are listed in order of most likely to comply with the Fourth Amendment to least likely to comply with the Fourth Amendment. These scenarios also anticipate that the MeCam will be capable of discretion and hiding while following and observing suspects. Although there is no reason to believe that the MeCam will have this capability, there is no reason to suspect it will not either. (And if this technology is later added by law enforcement agencies, courts may consider it a technological development that is not available to the general public.)

1. *Having spoken with several witnesses, municipal police believe they have identified a suspect that purchases bulk cocaine before storing, cutting, and selling it out of his home. The police obtain a judicial warrant to use the MeCam to follow the suspect into his home and record him intermittently over a two-week period. Using this evidence, the police obtain an arrest warrant and eventually a conviction.*

   Although this is clearly a search under the Fourth Amendment, because the police properly obtained a warrant based on reasonable suspicion, a court is highly unlikely to view this search as a violation of the suspect's Fourth Amendment rights.

2. *In response to complaints that an industrial manufacturer is willfully violating state environmental laws governing toxic waste disposal (which is a criminal offense), state police order a MeCam to investigate the plant. The MeCam flies and hovers throughout the property, recording activity. However, the drone decides not to enter a building or land on the property. The MeCam records employees loading deceptively labeled containers bearing toxic materials onto trucks underneath an overhang, which was undetectable by aircraft, although it could be seen from around the property. The trucks delivered the toxic materials to a facility that would not store such materials if it knew the actual contents of the containers. The police arrest the employees observed and their supervisors based on the MeCam's recordings.*

   Based on the existing cases addressing the Fourth Amendment and new technology, I suspect that courts will permit this police activity. The property is not a home, the MeCam is available to the general population, and the MeCam never even landed on the property or entered a building. The loading area, though shielded from aircraft, is within plain view from around the property, indicating that there was no intention of privacy by the manufacturer.

   But if courts ask whether the same activity would have been permitted if a police officer had entered the property, explored the exterior of the facility, and recorded employee activity, the constitutionality of the activity becomes more questionable. An officer could not have engaged in that activity without a warrant. Even though the MeCam represents technology that is readily available to the general public, courts should heed the Supreme Court's advice from *Jones* and consider the elements of AI surveillance drones that are unique from existing police technology. These elements include the ability to act, in some ways, more like a police officer than a police tool. An officer does not have to manually direct the AI into position. An officer does not have to tell it when to start and stop recording. An officer does not have to actively monitor it, either remotely or on the scene, for it to function. It would be prudent for courts to weigh that humanlike behavior when considering the constitutionality of the MeCam's actions. Assuming courts are willing to do that, it is more likely that a court will view the MeCam's investigation in this scenario as a violation of the Fourth Amendment.

3. *Based on rumors in town, municipal police suspect that the high school girls' junior varsity basketball coach is having an inappropriate relationship with one of his players. A detective assigns a MeCam to follow and record him and his car each day after school and practice for two weeks. The MeCam is equipped with advanced sound recording technology, permitting it to obtain more information about the coach than if it were bought off the shelf. The MeCam only follows the coach through public areas—roads, stores, town offices, etc. Using this information, the police arrest him for felonious sexual contact with a minor.*

Although the MeCam has followed the suspect through public spaces exclusively, a court is likely to view this investigation as a violation of the Fourth Amendment. First, this involves an individual, rather than a company, which brings it more naturally into the explicit authority of the Fourth Amendment. Second, the police upgraded the MeCam so that it used technology that was not generally available to the public. Third, the information recorded by the MeCam represents the scenario where individuals expect privacy for their cumulative public actions. An individual may do public errands, but does not expect those public errands to be chained together and mined for data about his or her life. This information, though based on public actions, retains an expectation of privacy that falls within Fourth Amendment protection, requiring either a warrant (that the police did not have) or reasonable cause (that is based on more than a town rumor).

However, there are other factors that a court could use to support the police department's use of the MeCam in this scenario. The MeCam is available to the general public. A human police officer performing the same investigation wouldn't need a warrant. And although the MeCam is cheaper than a human officer—removing some of the historical barriers to conducting this search—there are other barriers. The MeCam's battery life is likely somewhat limited. As the next chapter discusses, drones of that size typically have between 45 and 90 minutes of battery life. An investigation that uses the MeCam as the primary resource is likely to occur only in hour-long chunks at a time, limiting the intrusion into a person's private life.

4. *Based on a suspect's bad moustache, hemp clothing, and overuse of quotes from* Dazed and Confused, *a detective from the Mayberry Police Department sends a MeCam to enter his home through the chimney and use its best judgment to determine if there is anything there worth recording. The MeCam has been outfitted with special technology that is unavailable to the general public, permitting the drone to have a longer battery life and enhanced sensors to record more information about its surroundings. While there, the MeCam utilizes its enhanced sensors to read the suspect's diary, recording a passage in which he admits stealing nearly $10,000 worth of electronics merchandise from his job at Best Buy. Using this evidence, Mayberry PD arrests him and obtains a conviction.*

A court will declare this an unconstitutional search under the Fourth Amendment. It is in the suspect's home. There is no warrant. Even if the court decides to consider the search as if it were performed by a human police officer, this is clearly a violation of the Fourth Amendment.

Other Issues That Will Impact How Courts View AI under the Fourth Amendment

To a certain degree, courts will also look to the FAA's 2015 regulations for guidance. For example, in *Ciraolo* and *Riley*, the Supreme Court looked at FAA regulations governing permitted flight heights for planes and helicopters when ruling on the constitutionality of observations made from those aircraft. If the FAA defines airspace that it will consider suspicious, unlikely to support legitimate commercial behavior, dangerous, etc. for drones, that regulatory language will inform court decisions. Inasmuch as the FAA addresses AI drones in the 2015 regulations, courts will consider that language in its decisions.

Additionally, companies that develop and manufacture AI surveillance drones should draft analyses of the constitutionality of their products' uses. Companies can decide for themselves whether the documents should be available to the public. Some of the information that the analyses rely on may reveal sensitive information or protected trade secrets, which would discourage such availability. But these documents can inform courts about the thought processes behind the drones, creating a preemptive explanation (or defense) of the Fourth Amendment implications of the AI. Even if companies attempt to make them confidential, their public customers—municipal police departments and other government agencies—may be subject to Freedom of Information Act requests that could compel the release of the analyses. Either way, courts will consider these documents and may rely on them when considering the constitutionality of searches.

AI and Privacy

As AI surveillance drones become available to police departments, public concerns about what those devices mean for privacy and the Fourth Amendment will grow. What courts should be looking for is a way to protect people from Fourth Amendment violations while also providing government agencies with guidelines for using AI drones. AI drones will have legitimate uses for public safety, and those should be encouraged and protected, but not at the expense of unconstitutional government incursions into our privacy.

Although this chapter did not address other privacy concerns involving AI surveillance drones—Can individuals use them for surveillance? Where can individuals use them? etc.—those concerns are not a constitutional issue. Congress and state legislatures will have to address that type of privacy protection through legislation, as they have done through laws prohibiting stalking, public disclosure of private facts, intrusion of solitude, etc. Such laws would also have to comply with FAA regulations of aerial drones, so state legislators may want to consider laws that ban certain uses of (e.g., surveillance) and technology in (e.g., cameras) aerial drones while indicating that such bans are not intended to affect the rules and regulations promulgated by the FAA governing the *flight* of aerial drones.

As mentioned before, aerial drones also are an increasingly important part of international military operations, as well. Whereas domestic AI drone use must comply with the U.S. Constitution and American law, when the U.S. military uses aerial drones abroad, it must comply with international law. The next chapter discusses that topic.

CHAPTER 7

The Forthcoming United Nations Conventions on Artificial Intelligence

So far, we've essentially discussed American law exclusively. American law, like almost all domestic laws in a country, depends on the government enacting and executing laws. People living in countries with representative governments understand how these laws are supposed to work (current dysfunctional national governments excepted). Citizens vote for representatives who pass laws. Citizens vote (directly or indirectly) for presidents or prime ministers that execute and enforce the laws. If enough citizens dislike the laws or how they're enforced, they can elect new representatives and executives. Even people living in totalitarian nations have a basic understanding of how their laws function—the guys in charge do whatever they want—even if they have no official mechanism to change the laws they dislike.

However, citizens cannot "opt out." A nation's government always has the right to pass and enforce laws affecting all the people in that country whether all the people living in that country consent to being governed or not. Once a government is established, it has the right to govern. The best disgruntled citizens can hope for is to remove an unpopular government. They cannot simply inform the government "thanks, but we're going to find better laws."

In this chapter, we turn to an area of law where that model is turned on its head. International law is the system of laws dealing with the precepts and principles that govern nations in their mutual dealings and relations.[1] International law depends on customs and treaties, neither of which come from governments that have the absolute right to pass and enforce laws affecting nations.[2] Rather, they depend on the nations themselves to enter into the process of making international laws willingly and then act in good faith to enforce those laws within their borders and among themselves. In other words, each nation has to consent to being governed by international law, even when the potential outcome isn't what that nation wants.

But if it sounds like it isn't "real law," that's not the case at all. International law has no executive power that can compel compliance, as the president, governors, police, etc., do in America, but it does operate as law for nations. In general, they respect it and try to follow it. They do not always do so, but then again people don't always follow laws either: When was the last time you drove faster than the speed limit?

AI will impact international law. Like all other forms of law, international law assumes that all decisions are made by people, even decisions made on behalf of countries. In the last few decades, robots have begun performing tasks internationally that people always performed: minesweeping, navigating international air- and waterways, performing attacks in other nations, etc. But these robots are controlled by human beings. When a person decides to push a button to launch weapons at a foreign target from a drone, he engages in a potential act of war in the same way that a soldier who decides to fire a gun on a foreign target does.

When AI drones make the decision to launch an attack in another nation, that is something different. Can AI start a war? Can AI violate treaties? Can AI from two nations *form* a treaty? International law says nothing about AI or the effect of nonhuman decisions.

This chapter provides an overview of international law and how technology has changed it historically. I then discuss drone technology, before considering how international law may and should change if current drone use continues with AI drones.

International Law, or What Happens When Laws Are Purposefully Unenforceable

International law is a unique animal in legal circles because its laws are unenforceable by design. Inasmuch as the United Nations is sometimes derided as "world government," it doesn't have the enforcement power of an American town government. Smalltown, USA, can attach a lien to your house if you don't pay your property taxes, collecting when you sell the home or even seizing the property through a tax deed. If the United States refuses to pay its UN membership dues, the United Nations can remove its voting powers, but that's about it,[3] giving it the same sovereign power as the local Rotary. In some circles, the United Federation of Planets has more coercive force than the United Nations. Similarly, although there are numerous international courts—the International

Court of Justice, the International Criminal Court, the Permanent Court of Arbitration—none of them can force their jurisdiction; nations have to consent to their jurisdiction and then agree to follow each court's individual decision.[4]

Nonetheless, international law as it currently exists works fairly well in controlling the behavior of nations and providing processes for peaceful solutions to disagreements. Some of that can be attributed to avoiding embarrassment and loss of prestige among other countries.[5] But nations are also aware that at some point they are likely to need other states to consent to international law, even when they disagree with the outcome. So doing so themselves reinforces that helpful behavior.

A quick review of the historical development of international law will help illustrate how international law has emerged as an effective form of law, even without the power to coerce nations to comply. It will also show how new technologies have changed international law, suggesting how AI will do the same.

Early International Law

International law has a deep history, although in general it can be divided into pre-1648 and post-1648 periods, when the Treaty of Westphalia created the modern nation-state.[6] Almost as soon as civilizations developed, they became interested in relations with the outside world.[7] Early religious and secular writings reference peace treaties and alliances between different sovereign peoples: Jews, Romans, Syrians, Spartans, Carthaginians, German tribes, Arab tribes, etc.[8] The Roman Empire conducted extensive treaties with its neighbors, although the Romans frequently did this as a first step toward absorbing the other party into the empire or as formal recognition of the other party as a vassal to the empire. That is, there was no pretense of equality; the Romans were stronger and the treaties frequently admitted as much.[9]

Treaties recovered from early civilizations on Asia, Africa, and Europe show that the societies on those continents were concerned with many of the same topics simultaneously, but without interaction among many of the peoples. Ambassadors, extradition of criminals, protection of foreigners, and international contracts were all issues that early societies addressed through treaties. The principles that came to govern these matters are remarkably similar, regardless of whether the civilizations were in the Mediterranean region or on the Pacific Ocean.[10]

The Romans developed a concept that attempted to encapsulate these universal principles: *jus gentium*, the body of law that is common to all people. Long after the empire disintegrated and the concept of nations and states developed, *jus gentium* became one of the primary principles of international law, referring to universal law that has been consented to by all peoples and states throughout the world.[11] The combination of *jus gentium* and the patchwork of treaties among empires, tribes, and peoples constituted the general extent of international law prior to 1648.[12]

Modern Era of International Law

In 1648, the Treaty of Westphalia ended Europe's Thirty Years War. One of the main causes of the war—in addition to gold and general masculine preening—was the conflict between Catholic states and their Protestant minorities, and vice versa. The treaty provided safeguards for religious minorities, established equality between Protestant and Catholic nations, terminated the Holy Roman Empire, and recognized the sovereignty of its member states. This created the modern idea of a nation, one that has absolute control over the creation and enforcement of laws within its borders.[13] The international law that evolved from the Treaty of Westphalia addresses the relationships between these independent states.[14]

The basic sources for international law have not changed all that much since 1648, when treaties and commonly accepted customs were the basis for the rules that governed actions between nations. The International Court of Justice (ICJ)—which arbitrates disputes between nations and is the creation of the Charter of the United Nations, which itself was a multilateral treaty among 50 nations following World War II[15]—is viewed as a model for the consideration of sources for international law.[16] Article 38 of the Statute of the International Court of Justice states that the ICJ must make its decisions consistent with the following sources of international law, in order of preference, when making rulings to govern disputes between nations:

1. International conventions and treaties;
2. International customs;
3. General principles of law recognized by nations; and
4. Opinions of respected scholars regarding international law.[17]

Although Article 38 applies only to the ICJ by its terms, other international courts—the International Criminal Court, Inter-American Court of Human Rights, etc.—frequently rely on that list when making decisions regarding international law. Additionally, nations consider that list an authoritative statement on sources of international law as well.[18]

The list is consistent with the laws that have governed relations among nations, tribes, and peoples since the emergence of organized civilizations. Treaties are most important because they typically contain written rules that nations have explicitly agreed to,[19] but the customs and principles governing how nations conduct themselves internationally (the current form of the Roman *jus gentium*) are right behind those explicit agreements. (Like domestic law, international law considers guys writing books "nice to have, but not that important," dealing crippling blows to the egos of many authors who write about international law.)

Additionally, a few basic principles form the foundation of international law and support the treaties and customs that govern countries today. First, unlike Roman treaties that acknowledged power discrepancies, all nations are equal in international law.[20] This is not to say that all nations are equally rich or powerful, but international law treats all nations equally, in the same way that Americans insist that justice is blind and all people are equal in the eyes of the law. Second, the goal of the international legal system is for nations to do the most good to one another in times of peace and the least harm in times of war.[21]

It is worth discussing treaties and international customs individually, as the development of each since 1648 will impact how international law treats AI.

Treaties

Although there have been countless treaties among powerful, weak, and middling nations in the last 350 years, a few stand out as important. The Jay Treaty in 1794 between the United States and Great Britain is credited as being the first to establish international arbitration. The Jay Treaty resolved lingering disputes from the Revolutionary War that were not settled in the Treaty of Paris that ended the war in 1783. The treaty created commissions that would settle outstanding issues between the countries when they could not do so themselves through negotiations.[22]

Similarly, the United States objected to the United Kingdom's interference during the American Civil War, believing that country's actions violated its neutrality in an internal American matter. Under the Treaty of Washington in 1781, the United States and the United Kingdom agreed to submit American claims against the latter. When the United Kingdom began to comply with arbitration decisions beginning in 1872, it paved the way for greater arbitration agreements between nations, creating functioning mechanisms for enforcing international law between sovereign states.[23] In 1899, the Hague Peace Conference, a treaty between several European and Asian nations that other states later joined, created the Permanent Court of Arbitration, which gave nations a formal and agreed-upon process in international law (other than war) to use to resolve disputes among themselves. The Permanent Court of Arbitration, although still in existence today, was the forerunner of the International Court of Justice.[24]

In the 20th century, the Versailles Treaty that concluded World War I established the League of Nations, an earnest if disastrous attempt to bind the nations of the world into an international organization that would permit countries to work together on economic growth and human rights while peacefully settling disputes. The League was critically weakened even before it started, as the United States, which the organizing international leaders assumed would be an active participant, never even joined.[25] The United Nations, formed by the UN Charter in 1945, is the more successful successor to the League of Nations, containing 193 nations.[26] The United Nations uses a variety of tools to encourage actions from its member states, including "norm-creating" treaties and resolutions from the Security Council. These forms of UN action permit it to respond quickly to technological developments.[27]

For example, Resolution 1 of the UN General Assembly in 1946 responded to concerns related to atomic weapons, which were first used the year before. The resolution created the United Nations Atomic Energy Commission to address problems connected to atomic energy and permit its use only for peaceful purposes.[28] Subsequently, many nations signed multilateral treaties regulating and banning nuclear weapons, including the Treaty on the Non-Proliferation of Nuclear Weapons in 1970[29] and the Treaty Banning Nuclear Weapon Tests in the Atmosphere, in Outer Space and Under Water in 1963.[30]

Similarly, separate treaties have been signed by many nations to create agencies to govern certain areas of technology, but the agencies report to the United Nations. Examples include the International Atomic Energy Agency[31] and the International Telecommunications Union.[32] More recently, the Convention on the Prohibition of the Use, Stockpiling, Production and Transfer of Anti-Personnel Mines and on Their Destruction (the Land Mine Convention) became effective in 1999,[33] in response to the number of injuries and deaths caused by landmines, particularly in unstable areas of the world.[34] The convention, by its terms, relies on the United Nations for assistance in worldwide acceptance and compliance.[35]

The treaties and conventions addressing nuclear weapons and land mines are significant because they represent deliberate actions by the international community to change the customary international use of different weapon technologies. Similar treaties may be necessary for AI drones.

International Custom

Historically, custom governed much of international law, including conduct on the ocean, use of airspace and outer space, diplomatic immunity, and the rules of war.[36] Although there are now tens of thousands of treaties and international agreements, there is still room for custom to dictate international law. First, custom guides treaty interpretation. Second, treaties never bind all nations.[37] Take the Land Mine Convention, for example. As of early 2013, the United States was not a party to it.[38] Although the United States has not used land mines since 1991,[39] it is free to discontinue that policy and resume following customary use of land mines. This includes using land mines to reduce the mobility of enemy personnel, channeling them into specific areas or scattering them over a broad area. Nations use land mines to disrupt the formations and delay the movements of hostile forces. Additionally, minefields have been used to protect borders as a cost-effective solution to shortages of soldiers.[40]

In short, the way in which customary international law permits the United States to use land mines is the very reason why many people and nations believed the Land Mine Convention banning land mines altogether was necessary.[41] During the last 20 years of the 20th century, land mines claimed 1 million victims.[42] Somewhere between 75 and 85 percent of the victims were civilians.[43] Many of those were children. In some mine-infested areas, they are so desensitized to the danger that kids use the mines as wheels for toy trucks. Additionally, the countries most likely to be negatively affected are developing Third World nations that depend on agriculture for survival, where mines can destroy huge tracts of productive soil.[44]

Similar concerns about atomic weapons led to UN General Assembly Resolution 1 after World War II. Following the bombing of Hiroshima and Nagasaki, there was disagreement in the international community about the United States' use of atomic bombs. Many felt that the use was consistent with customary international law governing weapons and war; others felt that the atomic weapons rendered existing laws of war obsolete.[45]

By issuing a resolution and creating multilateral treaties governing atomic energy and nuclear weapons, most of the international community decided that the developing customs around atomic weapons were unacceptable and that different international law was necessary.

Similarly, most of the international community decided that the customary use of land mines had become unacceptable and drafted new law to govern it. It's likely that nations will make similar decisions about AI drones, as the current use of human-controlled drones has raised serious questions about their legitimacy under international law, particularly international humanitarian law.

Military Drones

Although use of sophisticated drones in military actions has grown exponentially in the 21st century, militaries have played with drones for many years.

Early Drones

There's a surprisingly long history of remote-controlled devices in war, possibly going back to mechanical pigeons in ancient Greece.[46] The true predecessors of today's military drones include unmanned surveillance balloons that dropped explosives[47] and kites equipped with cameras to take photos of enemy positions, both from the 19th century.[48] In World War II, the U.S. military used a variety of drones. British-born actor Reginald Denny formed the Radioplane Company and developed the Radioplane OQ-2. The U.S. Army and Navy bought 15,000 models for antiaircraft targets.[49] During the war, the American government also experimented with combat drone planes that could be operated remotely once they were in the air, but that needed a human pilot to get them off the ground before parachuting out. John F. Kennedy's older brother, Joseph, was one of the program's first pilots, but was killed when the drone's payload exploded prematurely.[50]

Late 20th-Century Development of Drones

During the decades following the Second World War, however, the technology began to improve significantly. Companies like Northrop Aircraft Incorporated and Ryan Aeronautical Company made significant improvements, permitting aerial drones to fly reconnaissance missions over North Vietnam, parts of China, and the Soviet Union. The drones did not operate in the battlefield, however.[51]

In the Yom Kippur War in 1973, Israeli air forces used unmanned drones to detect, and draw fire from, Syrian antiaircraft batteries, which permitted manned jets to attack the Syrians' position.[52] The success of the drones prompted Israel to develop a sophisticated system of intelligence-gathering drones by the end of the millennium.[53] The 1990s was an active time for the development of military drones, as advances in computing processing power and telecommunications permitted unmanned drones to have greater autonomy.[54] In 1994, the U.S. Air Force began experimenting with the Predator, a versatile aerial drone developed by General Atomics Aeronautical Systems in San Diego.[55] It operates remotely, but human personnel control it from another location, frequently on another continent.[56] In 1995, the Predator made its first combat deployment, conducting reconnaissance in the Balkans.[57]

Although the Predator made only minimal contributions to that conflict, officials within the Department of Defense were impressed by its real-time video of ground activity. This was a considerable improvement over relying on satellite images. During NATO's operations in Kosovo in 1999, the Predator distinguished itself as a combat tool. For example, while B-52 bombers attacked an area of southern Kosovo, a Predator circled Yugoslav troops beaming video to NATO military leaders of the effects of the attacks on those soldiers as it happened.[58]

Modern Military Drones

In 2000, the U.S. Congress passed a bill requiring one-third of all deep-strike aircraft to be unmanned by 2010 and one-third of all military ground vehicles to be unmanned by 2015.[59] In conjunction with that congressional directive, the terrorist attacks of September 11, 2001, pushed the world's lone superpower to begin a far-reaching campaign of attacks and surveillance using unmanned military drones.

Although the CIA had been operating unarmed drones over Afghanistan since 2000 for surveillance, following the September 11th attacks, the CIA began to arm the drones and assist in military operations. On February 4, 2002, an unmanned Predator drone committed a targeted killing for the first time.[60] The success of the Predator to provide real-time information to human soldiers while also conducting targeted attacks itself prompted the U.S. military to greatly expand the number of unmanned aerial drones they maintain. Prior to 2001, the U.S. Department of Defense deployed fewer than 50 unmanned aerial drones. By 2006, the number was over 3,000.[61] In 2012, the Pentagon used approximately 7,000 unmanned aerial drones.[62] At this point, the U.S. Air Force trains more drone pilots than conventional pilots.[63] Similarly, the number of unmanned ground vehicles increased from less than 100 in 2001 to nearly 4,400 in 2007 and 8,000 in 2011.[64]

Today, the Predator is one of many different types of aerial military drones that fall roughly into three categories: (1) micro and small, (2) medium altitude, and (3) high altitude, long endurance (HALE).[65] The micro and small aerial drones can be launched by hand or catapult, typically have a range of 3–10 miles, and a battery life of 45–90 minutes, although one of the largest models in this category, Scan Eagle, can fly more than 60 miles and has a battery life of more than 20 hours. They have a max altitude of 14,000 to 16,500 feet, although they typically operate at a fraction of that altitude—50 to 2,500 feet. These models are equipped with cameras and can be operated by soldiers on the ground to obtain real-time intelligence about enemy presence in their immediate area. Scan Eagle has autonomous technology that permits it to gather surveillance without direct human supervision.[66]

Medium-altitude aerial drones are capable of both reconnaissance and actual attack, depending on the payload. The Predator falls in this category. U.S. Air Force pilots remotely control or monitor them through one of three methods: direct controlled flight, semiautonomous monitored flight, and preprogrammed flight.[67] Medium altitude aerial drones operate at about the same altitude as a conventional commercial aircraft.[68]

HALE aerial drones are designed for wide-area long-term surveillance, acting like a "low Earth orbit satellite that's present all the time" over a field of operations.[69] They can fly at a higher altitude than medium-altitude aerial drones (65,000 feet versus 50,000) while conducting surveillance of over 60,000 square miles.[70] For reference, that means that only five HALE model Global Hawks are required to provide high-altitude surveillance of all of Afghanistan.[71]

AI Military Drones

All of the drones discussed above retain a "man in the loop," meaning that a human being has veto power over critical decisions like firing weapons or following a suspect off routine patrol paths.[72] Many models are capable of some autonomous activity. After a person provides the drone with directions (e.g., fly to these coordinates by this time and record activity there), the drone is able to make decisions on its own within the parameters of those directions (speed, altitude, route, etc.). Still, human beings with the military monitor these semiautonomous drones.[73]

The trend, though, is toward greater use of autonomous technology in military drones. Although numerous states claim that humans will remain "in the loop" on decisions to use lethal force, a UN report notes that "the advent of autonomous lethal robotic systems is well under way and that it is a simple matter of time before autonomous engagements of targets are present on the battlefield."[74] Israel is planning a "closed loop" border defense system, in which AI gun towers monitor the border, identify targets, and kill them without any human intervention. The Republic of Korea is developing an unmanned gun tower that can perform sentry duty along its border with the Democratic People's Republic of Korea. Although media reports indicate that human beings will determine when the tower may use lethal force, the tower will be equipped with the capacity to fire on its own.[75]

The U.S. military has publicly stated that for a significant period into the future, the decision to use lethal action will remain a human decision, but statements from high-ranking personnel suggest otherwise.[76] For example, Lt. General Rick Lynch stated in August 2011 that he was "an advocate of autonomous technology ... we have to continue to advocate for pursuit of autonomous technology."[77] Tests of autonomous attack drones have taken place at military bases like Fort Benning in Georgia.[78] On November 21, 2012, the U.S. Department of Defense issued Directive 3000.09, which attempts to establish department policy for the development of autonomous systems in drones.[79]

These signs point to the United States, the largest user of military drones, actively looking for greater AI in its military drones. However, inasmuch as existing drones have prompted concerns about their lawful use under current international law, AI drones will require wholesale changes to international law to accommodate combat decisions that are not made by human beings.

AI under International Law

As the previous section illustrates, although military drones have been in use for some time, it is only recently that they have become a direct and effective part of international combat. But they have also prompted questions about acceptable use of drones. Between 2009 and February 2012, drones killed between 282 and 535 civilians, including 60 children.[80] The Brookings Institution has estimated that drones kill 50 civilians for every militant target.[81] Although there are reasons to be skeptical of those exact numbers, the likelihood that drones have killed civilians due to the drone's mistake or the civilians' bad luck to be in the wrong place at the wrong time has prompted groups like the UN Human Rights Council to question whether the use of drones violates international humanitarian law.[82]

The use of drones internationally has prompted a great deal of analysis, questions, and hand-wringing among nations and international lawyers.

Although experts in the field debate whether drones violate or are consistent with international legal norms, it will be helpful here to focus on a particular area of international law: law of war, or international humanitarian law. Books and articles have already identified this as an area of law where there are serious questions about the use of drones operated by human beings.

Those devices, however, are still controlled by human beings, with human beings directly responsible for their actions. As Philip Alston, the UN Special Rapporteur on Extrajudicial Executions, noted, "a missile fired from a drone is no different from any other commonly used weapon, including a gun fired by a soldier."[83] AI drones are different because there will be no soldier or pilot pulling the trigger; a machine will make the decision. Whereas international law might be ambiguous now regarding drones, it is mostly silent on AI drones right now. Even when international law contemplates other AI, like autonomous cars and planes, it prohibits their use.

AI Drones and International Humanitarian Law

International humanitarian law (at times called law of armed conflict or the law of war) applies to situations of armed conflict and governs the conduct of hostilities and the protection of persons during conflict.[84] At its foundation, international humanitarian law is based on four principles:

1. All parties to a conflict must distinguish between individuals who are fighting and those who are not, directing attacks only at those who are.
2. All parties to a conflict must seek to minimize incidental casualties during war, meaning that the means and methods of attacking an enemy are not unlimited.
3. A military has the right to use any measures not forbidden by the laws of war that are indispensable for securing the complete submission of the enemy as soon as possible.
4. All parties must seek to minimize suffering in an armed conflict.[85]

Two kinds of weapons are prohibited in armed conflicts: indiscriminate weapons[86] and weapons that cause unnecessary suffering.[87]

With regard to indiscriminate weapons, under Article 51(4) of the Protocol Additional to the Geneva Conventions of 12 August 1949, and relating to the Protection of Victims of International Armed Conflicts (Additional Protocol I), indiscriminate weapons are ones that produce attacks that are: 1) not directed at a specific military objective; 2) employ a method or means of combat that cannot be directed at a military objective; or 3) employ a method or means of combat that produces effects that cannot be limited as required by Additional Protocol I. Put more succinctly, indiscriminate weapons are "weapons that are incapable of distinguishing between civilian and military targets."[88]

With regard to weapons that cause unnecessary suffering, the International Court of Justice has stated that "it is prohibited to cause unnecessary suffering to combatants: it is accordingly prohibited to use weapons causing them such harm or uselessly aggravating their suffering . . . States do not have unlimited freedom of choice of means in the weapons they use."[89] The main idea behind this international law is that weapons which increase suffering without increasing military advantages in any way are unlawful.[90]

There is some disagreement as to whether remotely controlled drones violate either of these prohibitions. Some scholars and military leaders describe these drones as armed robotic "killers"[91] that separate the killer from the victim, making atrocities more likely.[92] In this light, drones permit more destruction than any other weapons system and become tools for indiscriminate attacks and unnecessary suffering.[93] However, others argue that current drones use the same weapons carried by piloted fighter aircraft, which are not banned by any international agreement. Some writers press the point, noting that because drones can conduct surveillance before making an attack, they actually reduce indiscriminate attacks as defined by Additional Protocol I and help to eliminate unnecessary suffering because they can track a target for hours and strike when civilians are not in the area.[94]

Supporters of drones also point to the foundational principles of international humanitarian law as evidence that the use of remotely controlled aerial drones is permitted. In addressing the first principle—that parties to a conflict must distinguish between individuals who are fighting and those who are not, directing attacks only at those who are—Article 48 of Additional Protocol I states: "In order to ensure respect for and protection of the civilian population and civilian objects, the Parties to the conflict shall at all times distinguish between the civilian population and combatants and between civilian objects and military objectives and accordingly shall direct their operations only against military objectives." If drones permit remote human operators to conduct enhanced and sophisticated surveillance, it is possible that they can discern civilians from military personnel better than soldiers on the ground.[95]

Similarly, while international humanitarian law does not prohibit civilian deaths, it does require that a nation's objective in an armed conflict be to weaken the military forces of its enemy, not its civilian population.[96] Drone surveillance technology permits human operators to determine the civilian presence in the surrounding area of an attack, possibly letting them understand better than soldiers in the field what opposing military force is present. Drones may be better equipped to weaken an enemy's military while inflicting as little harm as possible to civilians, which promotes the second principle, that parties to a conflict must seek to minimize incidental casualties during war.[97]

The crux of the argument between those who believe the current use of drones violates international humanitarian law and those who believe it is consistent with that law is whether or not drones operated remotely by human beings constitute an entirely new weapon or merely a more precise version of existing lawful weapons. If drones are a new weapon, they could represent an indiscriminate weapon that causes unnecessary suffering. If drones are merely Existing Lawful Weapons v. 2.0, international law has already determined that they are fine under international humanitarian law.

This argument will be irrelevant when nations start using AI drones. Even a rifle when operated by an autonomous soldier becomes a new weapon because a machine will survey the human beings in its range, sort out civilians from enemies, and decide to open fire. Numerous scholars rely on the reasoning of the UN Special Rapporteur on Extrajudicial Executions, Philip Alston, to describe human-controlled drones as "no different from any other commonly used weapon" because the "critical legal question is the same for each weapon: whether its specific use complies with" international humanitarian law.[98] As long as a human being makes the decision to fire, the use will be the same whether it is a human soldier in the field pulling the trigger or a human operator on the other side of the planet pushing the drone's missile launch button.

International law, by its terms, does not effectively address AI and machine-made decisions. Consider Article 48 of Additional Protocol I, discussed above. It only directs human beings unambiguously: "The *Parties* to the conflict shall at all times distinguish between the civilian population and combatants" (emphasis added). The Parties are not people, of course, but they are nations that are run by people, with people carrying out national decisions in fields of combat. That is the assumption in international humanitarian law, as embodied in Article 91 of Additional Protocol I: "A Party to the conflict . . . shall be responsible for all acts committed by *persons forming part of its armed forces*" (emphasis added). Decisions made by human soldiers are accounted for in international humanitarian law.

But when a nation sends an AI drone into a conflict, there is no accountability, not by the letter of the law. If an AI drone acts disproportionately to the military objective or kills civilians, the country responsible is *not* responsible. Its leaders can plausibly say: "While we acknowledge that the AI drone is ours, we do not condone its actions. Although our military personnel were overseeing its operations, by its design it makes independent decisions about its assigned objectives. Usually those decisions are very good. Today the decisions were bad. However, we acknowledge no responsibility for those decisions, as the persons of our armed forces did not make them."

Some policy and non-profit groups, like Human Rights Watch, point to this potentially lethal error as proof that international humanitarian law prohibits AI drones, noting that armed forces must distinguish between combatants and noncombatants.[99] But acts and decisions of armed forces under international law, as noted in Article 91 of Additional Protocol I, seem to include only those acts performed and decisions made by people, not machines. Even if we were to contort Additional Protocol I to include AI drones, it's not clear what is required of nations. Is it that AI drones can *never* fail to distinguish between combatants or noncombatants? That standard seems unlikely, as human soldiers would fail that strict test. The U.S. Department of Defense in Directive 3000.09 seems to assume that if there is a burden, it is only to create procedures to test and oversee the operation of AI drones to ensure that the technology is capable of distinguishing among targets successfully, not that it does so all the time.[100] In its silence on AI, international law permits almost all legal interpretations of its application to AI.

That is the issue regarding the liability of parties who use AI drones that violate international law. To be sure, Article 91 does not explicitly state that a Party to a conflict is *not* responsible for acts committed by AI, but international humanitarian law says nothing as explicit about AI as it says about humans. In fact, it says essentially nothing about AI. At best, international humanitarian law inadvertently addresses AI. For example, nations are required to take "constant care" and precautionary measures to ensure that civilians are not injured during attacks.[101] The standard of precautionary measures is "whether a reasonably well-informed person in the circumstances of the actual *perpetrator*, making reasonable use of the information available to him or her, could have expected excessive civilian casualties to result from the attack" (emphasis added).[102] The International Criminal Tribunal for the Former Yugoslavia, which issued that statement of law, clearly intended this standard to be for a *human* perpetrator, but it could arguably be contorted to apply to an AI drone as well.

But then the question becomes what information was available to the AI. It may be harder to appreciate the information that was available to AI than the information that was available to human beings in the same circumstance. A human being likely had people around him or her—other people in a crowd, subordinates, friends, superiors, etc. A human being had a perspective that other human beings can relate to and use to analyze his or her decision. But an AI drone likely acted alone. An AI drone may have monitored the site of the killing from 10,000 feet in the air. An AI drone's error may be due to the AI or a human commander.[103] The result of those errors may be that both the AI and human took appropriate precautionary measures to protect civilians. There is a maze of complexity in these factors that may make it impossible to properly assign liability and guilt under current international humanitarian law.[104]

Much like autonomous cars and "operator" liability, one train of thought on AI drones suggests that if we assume human control at all times, then autonomous drones comply with international humanitarian law because a person is always in command of them, like any other weapon. And it is true that if an AI drone were programmed to complete a battlefield mission autonomously while under constant human monitoring, it could still be described as under human control.[105] But as described above, under current international law, this is an ambiguous prospect at best. International law does not state that human operators are responsible for their "thinking weapons" because until recently thinking weapons were not a reality.

Similar to issues we've discussed with other AI, there is evidence to suggest that it may not be realistic to expect humans to retain the same level of control over AI drones as they have over existing conventional weapons. First, people controlling drones tend to work long shifts—up to 12 hours—and the boredom associated with watching a device that can function independently (think about monitoring your washing machine) seriously impacts a person's ability to successfully monitor a drone.[106]

Second, communications between drones and their human operators still have significant issues. The accident rate for aerial drones is far above that of manned aircraft, caused at least in part by disrupted communications between drone and ground control.[107] Additionally, there is a 2–5 second lag time between the time when the drone records video and when those images appear on the operators' screens, leading to potentially fatal accidents, like when a child walks out from behind a building under surveillance instead of the expected enemy combatant.[108]

Because drones are currently only semiautonomous at most, and the decision to launch an attack is made by a human operator, those communications issues have not resulted in an effectively unmonitored drone launching an attack. With AI drones, problems with communications will be much more likely to lead to that outcome. Twisting current international norms to apply to AI drones by assuming a human is always in control of an autonomous weapon distorts what actually happens.

UN Conventions on Autonomous Drones and Artificial Intelligence

This absence of law is not unique to international humanitarian law. A review of almost every area of international law that will be impacted by AI—national sovereignty, law of the seas, targeted killings, etc.— reveals that none of them address AI directly. Any opinions about AI under international law are based on expected customary uses by nations and interpretations of existing international law that twist the intended meaning. Obviously, this is presently a nonissue. AI is not used in drones and is not yet functional enough in general to force changes in international law. But the time is coming shortly when there will be AI in drones, and they will function in areas that impact international law.

The best way to resolve disagreements, potential issues, and conflicts between nations over AI is to address them in a multilateral treaty, maybe one organized by the United Nations. AI presents tremendous benefits and dangers, similar to atomic energy. Following World War II, atomic energy had the potential for great energy, but also great destruction. AI holds similar promise. AI can save human beings from hard labor and improve the quality of their lives. But AI also permits war to be waged like video games, disconnecting soldiers from violence and preventing them from realizing the truth of Robert E. Lee's observation that, "It is well that war is so terrible, otherwise we should grow too fond of it."

But the benefits are endangered, and the threats are greater, when there is international confusion about what use of AI is acceptable and what use is not. Even though there is no executive office to enforce international law, it provides a useful road map for nations: "We definitely can do this, but we should think carefully about doing that other thing." It's like the line at an ATM. Although there is no enforceable law telling the second person in line how far back he has to stand from the first person in line, there is general agreement that you should stand back a few feet. Everyone is more comfortable when that rule is obeyed, helping maintain order in the ATM vestibule. Similarly, when there is general agreement about international laws, all nations are more comfortable and there is less conflict.

The greatest obstacle to a UN Convention on Artificial Intelligence is the cooperation and acceptance of the nations that will have started using and relying on AI. Nations that already rely on drones—particularly the United States—will likely be early adopters of AI drones. Those nations—postindustrial countries with the economic incentive to research and develop labor-saving technologies—are also the most likely to adopt AI in other forms that impact international laws: self-piloting planes, self-sailing ships, etc. The temptation in a convention on AI is to draft explicit guidelines for everything: AI cannot open fire on any human beings, AI cannot enter another country without express permission from that country, nations are liable for the acts of their AI, etc.

But some of those bright-line tests will be unacceptable to the United States, Israel, and other countries that will have already started using AI. If their use of existing drones is any indication, they will use AI drones in similar ways and will be opposed to a treaty that bars that use. Those nations will not sign the treaty, and it will be much less useful.

Rather, I suggest multiple conventions that will address different areas of AI use with different degrees of specificity, permitting the United States to sign and ratify some but not all of them. This is not ideal, but it would provide some internationally accepted norms about AI and create defined aspirational goals for AI use under international law. In many ways, the Land Mine Convention serves this aspirational function today, as some of the most relevant countries (like the United States) have not signed, but it has created an ideal status for antipersonnel land mines under international law: no nation will use, produce, or stockpile antipersonnel land mines.[109] This ideal has effectively stigmatized the use of antipersonnel land mines. By 2010, production of antipersonnel mines had ceased in 39 nations, 5 of which are not parties to the Land Mine Convention, and legal trade in the weapon is virtually nonexistent.[110]

Although numerous configurations for the AI conventions are possible, I recommend the following:

- *Convention on Artificial Intelligence and Liability of Nations*: This treaty would affirm that nations are liable for the actions of their AI as they are for the actions of their military personnel. The articles should address precautionary measures required of nations while operating AI. The intent of that would be to provide clear guidelines for states to follow when developing and using AI. The main point of this convention would be to give AI a status equivalent to humans under international law in terms of creating liability for their nations, while also recognizing that some of the precautionary measures that nations need to maintain when operating AI are different than the precautionary measures they maintain with their human forces. For example, training requirements for AI drones and the humans who oversee them will be different than the training requirements for soldiers who work with traditional conventional weapons.

- *Convention on Artificial Intelligence and the Sovereignty of Nations*: Although this topic will cause a fair amount of disagreement among countries, there should be a treaty that lays down rules governing AI and the sovereignty of nations. In the same way that there are questions about a drone conducting targeted killings in another nation, there will be even more confusion about AI entering the airspace of another country. A bright-line rule prohibiting a nation's AI from violating another nation's sovereign territory will be rejected by the states that already use AI drones. But a convention that incorporates the right of each nation to self-defense[111] and recognizes that there are instances when such violations are necessary (or at least too convenient to pass up) will find more supporters. Such an international standard will also more accurately reflect the reality of how nations use AI.

- *Convention on Artificial Intelligence and Self-Piloting Vehicles*: In order to facilitate the integration of AI drivers and pilots into international trade and travel, it would be helpful to have agreed-upon standards for AI pilots of air, ground, and water vehicles. Nations need to know what their responsibilities are for AI sea vessels registered in that country, AI warships navigating international waters, and AI ships responding to other vessels in distress, among other issues. Nations hoping to use autonomous planes and ground vehicles have other challenges.

  In addition to the creation of new international laws governing AI planes and cars, nations hoping to embrace pilotless planes and driverless cars must revise or overturn existing treaties that address self-flying planes and self-driving cars. The Convention on International Civil Aviation specifically prohibits a "pilotless aircraft" from being flown over a nation without the specific authorization of that nation.[112] A treaty providing such authorization at large would adjust international law to give pilotless planes an assumption of legality, while also providing nations with rules for using pilotless planes.

  Similarly, countries looking to embrace driverless cars and trucks will need more than new domestic legislation to provide guidelines for their citizens. They will need new treaties too, as the Geneva Convention on Road Traffic requires that every vehicle have a driver who drives in a reasonable and prudent manner.[113] Countries will need to revise those provisions of international law if they want to fully embrace the potential of autonomous cars to drop off owners at work before driving home and ferrying kids to school. Countries like the United States, where autonomous cars are already appearing, should consider revisiting this sooner rather than later.

- *Convention on Artificial Intelligence and Intellectual Property*: The creation of intellectual property by AI is more thoroughly discussed in the next chapter, but international standards will be important as machines, robots, and programs begin to create more content. When AI writes a best-selling novel, the duration and terms of the copyright should have some consistency from nation to nation.

- *Convention on Artificial Intelligence in Surveillance*: The drones currently available are capable of monitoring our activities in ways that were unimaginable just a decade or two ago: collecting real-time video, watching one geographic location for hours at a time, and hiding in small places to observe people on the ground. The surveillance technology will continue to improve, while drones will begin to incorporate AI, permitting them greater flexibility in their monitoring operations. Without the need for direct human control, AI drones will be able to respond to their sensors, redirecting themselves automatically to areas or people that appear to represent the most relevant observation opportunities. They might even develop an intuition of sorts. With this technology in mind, it will be important for states to address basic questions about surveillance AI drones. When are they authorized? When must other nations be notified? Are there limits on the types of people drones may monitor? How close to people may AI drones operate? A multilateral treaty can address all of these questions at once, although convincing states already using AI drones to limit their use may prove difficult.

- *Convention on Artificial Intelligence in Armed Conflict*: This treaty would expand international humanitarian law to address mandatory conduct of AI during military actions. It would govern the military use of AI against enemies, when civilians are in danger, in targeted killings, and other similar contexts. Depending on the negotiated terms, it could provide a standard for nations to use when determining if AI drones are authorized under international law to operate and attack, in which case countries that already use AI drones may decide to sign. However, if the treaty prohibits any attacks by AI drones, as groups like Human Rights Watch demand, those nations are likely to reject it altogether. Either could represent an unfortunate aspirational treaty, as countries like the United States may decide that they do not want to limit their military use of AI drones in *any* way.

These suggested treaties govern a wide range of AI uses and issues. I believe that most nations will agree with most of the outlined conventions. Agreements regarding liability, intellectual property, and self-piloting vessels should find greater consensus than agreements governing sovereignty, surveillance, and armed conflict. AI drones will have many military and espionage uses, just as human-operated drones do now. Nations will be reluctant to give up that utility, particularly when they get into the prisoner's dilemma of trying to figure out if their rival nations will also swear off AI drones. No country will be eager to tie itself to that commitment when other countries are not limiting themselves that way.

International AI Standards

Even if all of the relevant countries are not willing to renounce the more controversial uses of AI, creating standards for AI in international law is useful. It provides a model that the noncomplying nations can look to for guidance when making decisions about AI. For example, even though the United States is not a party to the Land Mine convention, the military is aware of its goals, and that affects its approach to antipersonnel land mines. Conventions governing AI and armed conflicts, surveillance, and sovereignty can work the same way. Additionally, treaties addressing less controversial aspects of AI will provide a solid framework for the use of AI in international relations.

AI is poised to radically change many fundamental elements of our lives and economies. Workforces will be displaced. Transportation will be easier. Production will be cheaper. It is important that the many nations on earth approach AI with some fundamental concepts in common. Hopefully, this will mitigate some of the more harmful elements of AI, like the loss of many jobs, and promote some of the benefits of AI, like greater freedom for people to become better educated and create new jobs, businesses, and technologies. This will facilitate the expansion of the benefits of AI to more people in more countries.

One of the benefits may be the development of AI that can create—AI that writes, composes, and draws. This type of AI may be the least developed of any discussed in this book, but you can see its beginnings already. The next chapter discusses what happens when AI creates intellectual property.

Part III

WILL A ROBOT PROTECT ITSELF?

CHAPTER 8

What Does a Robot Own?

Of all the AI topics addressed in this book, this chapter addresses the development that is most promising, that is most threatening, but that is also the furthest from being a reality. AI that creates original ideas could revolutionize the world ... or it could destroy the last quality jobs that humans alone are qualified to fill. The answer lies somewhere in the middle, hopefully closer to the former than the latter.

AI that can create original ideas could expedite the creation of new inventions, further our understanding of the universe, develop solutions to the worldly problems we've long struggled with, and provide endless hours of books, movies, and music for our entertainment. That, at least, is the potential. If AI were to reach its potential, though, it would threaten to displace human inventors, astrophysicists, scientists, authors, actors, and musicians in the same way that industrial robots displaced human industrial workers last century. Threatening, to say the least.

Our intellectual property system is designed with only human inventors and authors in mind. It doesn't consider nonhuman creators. If a robot creates a new processor or engine, is it entitled to the patent for those inventions? If a robot writes a book, is it the recognized author of that book and entitled to the copyright? A robot that can create forces these questions and a more fundamental one: When a robot creates intellectual property, is the robot entitled to ownership of that intellectual property or is the human who created the robot entitled to ownership? Currently, our laws do not address this at all.

In the first chapter, we discussed in broad terms intellectual property created by AI, looking specifically at Siri and what happens if Siri makes profitable media. This chapter takes a closer look at this topic, highlighting other AI creators—programs, robots, and devices that either exist or are in development and are expected to create profitable new media, new inventions, and new ideas. Having reviewed those forms of AI, I will return to Ralph D. Clifford's examination of whether AI can own IP and suggest ways that our intellectual property laws should be amended to incorporate AI into our legal system while also spreading its benefits to as many people as possible.

Brief History of Intellectual Property

As with other chapters, though, it is helpful to have a quick review of intellectual property law, where it comes from, why we have it, and what it protects.

Overview of Intellectual Property Classifications, Definitions, and Justification

Intellectual property refers to the legal rights that result from intellectual activity in the industrial, scientific, literary, and artistic fields. Generally, there are two types of intellectual property: industrial property and copyright. Industrial property includes patents (essentially a monopoly on manufacturing, using, selling, or importing an invention[1]), industrial designs (design for product appearance and packaging[2]), and trademarks (a sign or mark that distinguishes the maker of that product from other makers[3]). Copyright, which protects an owner's monopoly to copy a creative work,[4] includes literary, artistic, and scientific works. We protect intellectual property rights primarily for two reasons. The first is to protect the moral and economic rights of creators. The second is to promote creativity to contribute to economic development.[5]

Early Development of Intellectual Property

The idea of protecting intellectual activity and creation has deep roots. Noted intellectual property historian Abraham S. Greenberg cheekily noted that even God wanted to mark his/her creation in the book of Genesis, as he applied a sign on Cain to note that he was made by God.[6] The ancient Greeks used an awards system to recognize design achievements, which performed some of the same functions as the modern patent system.[7] Their pottery, sculptures, and other manufactured goods had symbols on them to note the tradesman who created them,[8] which are the predecessors of the Nike swoosh and other modern trademarks. Similar marks and symbols functioning as trademarks have been found on Chinese pottery, possibly dating as far back as 2698 BCE,[9] and in many other ancient societies, including the Egyptians, the Vedic civilization (which flourished in the northern area of modern-day India), and Assyrians.[10]

The concept of intellectual property continued to develop during the Roman period. The Roman Empire had an incredible variety of trademarks.[11] Roman potters alone used approximately 6,000 trademarks.[12] Additionally, Roman authors had a sense that their intellectual creations were valuable, as they complained about the exploitation of those creations.[13] Their sense of injustice was probably heightened by the fact that there were laws and traditions in place that supported their belief that only *they* could exploit their creations. Roman authors could, in fact, make money from the copying, duplication, and publishing of their works because the monetary value of their intellectual creations was recognized.[14]

The Middles Ages and England's Contributions

However, it appears that this sense of copyright protection was lost in Europe during the Middle Ages. During that period, most works of art were considered part of the public domain, meaning anyone could copy them without paying the author.[15] Much of this has to do with the Catholic Church, which was by far the largest sponsor of art—music, paintings, sculpture, architecture, etc.—and therefore the owner.[16] The Church heavily censored content in order to enforce belief, and permitting widespread use of its art and literature—rather than "profane" art—promoted this goal.[17]

It is fair to say that the development of the moveable-type printing press radically changed Western Civilization's concept of this type of intellectual property, to the point that some mistakenly believe that intellectual property began at that time.[18] With that technological advancement, it became easier to print and obtain books that did not have the Church's blessing. It also became easier to pirate books, reselling them for an unearned profit. But there was no universal protection in Europe for literary output for several hundred years after Johannes Gutenberg introduced the moveable-type printing press to the continent.[19]

During that time, there was, instead, a "privilege" system in which a ruler granted certain printers the privilege of printing books, but for no longer than five years. There was no system to protect authors.[20] Rather, authors could petition rulers for protection after their works were copied and hope that the ruler would be sympathetic; there was no guaranteed protection from pirating copyright material.[21] It was not until 1709, when Great Britain's Parliament passed the Statute of Anne, that copyright first became a regulated form of intellectual property, protecting the author. Under the Statute of Anne, authors had a monopoly over their works for 28 years after first publication.[22]

England in the Middle Ages and the Renaissance played a pivotal role in the development of patents as well, as few governments prior to that time had focused so specifically on their manufacturing sectors. By comparison, the Romans were not an industrialized society. They were more concerned with trade than with production, and their laws had little to say explicitly about the protection of patents and inventions.[23] (In contrast, Roman law had provisions to protect proprietary information and trade secrets.[24])

The idea of patents appears to have originated in early European commerce, when monarchs granted exclusive privileges to certain commercial ventures as a way to provide protection to individuals, guilds, and cities that undertook business ventures risky to life and wealth.[25] English kings took this a step further in the early 14th and 15th centuries, granting monopolies and special privileges to merchants who brought a new technology or technique into England.[26] The English crown initially established this policy because British industrial and commercial development lagged behind most of continental Europe. This practice was successful, leading to the development of the British cloth industry, the country's first considerable manufacturing industry, and attracting producers of salt, armorers, shipwrights, glassmakers, and ironworkers. Queen Elizabeth altered the practice to encourage domestic British industry, creating the system of protecting invention that would become the modern patent system.[27]

International Efforts to Protect Intellectual Property

In the 1800s, international efforts began to bring uniformity to intellectual property laws across borders. In the second half of that century, entrepreneurs and authors began to work together to protect their interests. Prior to this time, many people in Europe and the United States believed that intellectual property laws within a country rightly protected authors and inventors from that country, but that authors and inventors in other countries did not need protection in other nations. Rather, taking better ideas from other countries contributed to the advancement of mankind and the dissemination of knowledge.[28] So, for example, many American publishers saw no legal or moral hurdles to publishing

Charles Dickens's works in the United States; he was British and it was just (and profitable) to share his books with American audiences regardless of whether he authorized that publisher to do so. As you might expect, this problem was particularly troublesome between countries that shared a language, but where one had a distinct development advantage (like 19th-century America and England; Belgium and France).[29] This began to change with the Paris Convention for the Protection of Industrial Property of 1883 and the Berne Convention for the Protection of Literary and Artistic Works of 1886.

The Paris Convention gave foreign applicants for industrial intellectual property rights (patent, trademark, etc.) the same rights as a citizen of that country applying for the same industrial intellectual property rights.[30] A Russian inventor applying for a patent in Germany must receive the same rights as if a German were applying for the same patent. The Paris Convention recognizes a broad array of intellectual property, relating to industrial and commercial property, but also to agricultural and extractive industries and to the manufactured and natural products industries (e.g., wines, grain, tobacco, cattle, minerals, etc.).[31]

The Berne Convention requires countries to grant the same copyright to authors from other countries as they grant to their own authors.[32] So Italian copyright law applies to anything published in Italy, regardless of the nationality of the author. This protection is available to literature and scientific writings in all forms, dramatic or musical works, choreography, cinematographic works, maps, three-dimensional re-creations, illustrations, painting, architecture, sculpture, photography, and other forms of art.[33]

Both treaties—along with others governing further areas of intellectual property, like trademark,[34] integrated circuits,[35] sound recordings,[36] etc.—are still in effect today. Both treaties also apply exclusively to people. The Paris Convention refers frequently to "nationals," clearly intending that they are "persons" from a country.[37] Similarly, the Berne Convention uses the term "authors" and indicates that they are only people.[38]

International efforts to promote laws governing intellectual property law have continued as technology has developed. For example, as computers, the Internet, and digital media became much more widely used in the 1990s, the World Intellectual Property Organization organized the WIPO Copyright Treaty of 1996. It explicitly extends copyright protection to computer programs and databases.[39] It also states that authors retain the exclusive right to authorize any communication to the public of their works (i.e., publication or release), regardless of whether that publication or release is done "by wire or wireless means."[40] This language is intended to give authors control over all releases of their work made over the Internet.[41]

American Intellectual Property Laws

American intellectual property law is mostly consistent with these international efforts, as the United States is a party to many of the treaties governing intellectual property across national borders. However, I want to draw attention to American laws governing patents and copyright, as those are likely to be the most important when AI begins producing original content.

The U.S. Constitution grants Congress the power to protect intellectual property, by empowering it "To promote the Progress of Science and useful Arts, by securing for limited Times to Authors and Inventors the exclusive Right to their respective Writings and Discoveries."[42] As chapter 1 stated, U.S. copyright protection "vests initially in the author or authors of the work."[43] Authors generally retain exclusive rights to use their works during their lifetimes; however, the protection exists for 70 years after their deaths.[44] Copyright protection for anonymous works, pseudonymous works, and works done for hire (think the Disney catalog of characters) lasts for 95 years after publication, or 120 years from the year of its creation (whichever expires first).[45] This represents a serious expansion of copyright protection over the last 220 years. Congress only granted up to 28 years of copyright protection when it passed the Copyright Act of 1790.[46] The language of the act is largely the same as the Statute of Anne.[47] The expansion occurred gradually, as the Copyright Act of 1831 extended the protection up to 42 years, the Copyright Act of 1909 extended the protection up to 56 years,[48] and the Copyright Act of 1976 extended protection to either the life of the author plus 50 years (for works with authors) or 75 years (for anonymous works, pseudonymous works, and works done for hire).[49]

An inventor or developer seeking an American patent may apply to the U.S. Patent and Trademark Office (PTO) for a patent if he or she "discovers any new and useful process, machine, manufacture, or composition of matter, or any new improvement thereof."[50] Inventors *must* apply to the PTO for a patent; unlike copyright, they do not vest automatically. Filing with the PTO is important. If two people develop the same invention, the first one to file will receive a patent.[51] A patent creates a monopoly for 20 years in the invention or development.[52]

Like the Paris and Berne Conventions, U.S. intellectual property law assumes that people invent and create. The Constitution refers to

"Inventors" and "Authors." An inventor is typically defined as "a person who invents";[53] an author is typically defined as "a person who composes a book, article, or other written work."[54] U.S. statutes and code do not attempt to define those words more broadly, either. Patent law refers to "patentees," but notes that the term includes those "to whom" the patent was issued.[55] Patent law also refers to inventors, but defines them as "individuals" who invented or discovered the subject of the invention.[56] Similarly, copyright law, as alluded to in chapter 1, clearly only intends to protect intellectual property created by people. Although this is not stated outright, there are numerous provisions that demonstrate it nonetheless. For example, whenever an author dies, the copyright shall be transferred to the remaining spouse or children. The only apparent exception to this is for work made for hire, as the employer owns the work, not the author.[57] Similarly, although author is not defined, "anonymous work" is defined as lacking a "natural person" (i.e., an actual person and not a company) as the author.[58] Although that suggests that any work created by something other than a living, breathing person is considered an anonymous work under U.S. law, anonymous works are protected in order to permit the actual author to reap the benefits of the copyright later if that person is discovered, meaning AI-created works cannot qualify as anonymous works.[59]

AI That Creates

Although I mentioned a few forms of creative AI in the first chapter, I want to give that area of AI a more thorough review. Recreating human imagination represents a form of weak AI that is arguably just short of strong AI like the title character in Isaac Asimov and Robert Silverberg's novel, *The Positronic Man*. The ability to imagine and create is a fundamental part of what makes us human. But creative AI is already here, having developed quietly over the last several decades. It cannot perform in the full array of human arts yet, but it's getting closer.

Music

Although Emily Howell's 2010 CD, *From Darkness, Light*, was a remarkable achievement of AI composition, it did not develop from a vacuum. Emily represents nearly three decades of research and development by David Cope. Experiencing writer's block while working on an opera in 1980, he wrote a computer program that "could create some kind of drivel that would inspire" him to compose again.[60] At the time, computers were so new that they only made noises "when you booted them or made a mistake," and he had to use number variables for pitch, duration, and volume, which he then translated into printed music he could play.[61] He wanted to create virtual David Cope software that could write new pieces in his style.[62]

Turns out, getting a computer to compose like a person is tricky. He spent a year studying artificial intelligence and programming, trying to create a rules-based program that could replicate his taste. When that proved too difficult, he scaled down his ambitions to a program that could write chorales, four-part vocal hymns, in the style of Johann Sebastian Bach. He eventually produced software that could compose chorales "at the level of a C-student college sophomore," meaning he'd given birth to a technological marvel and a musical idiot.[63] It was a rules-based program that could not compose anything full of life or capable of surprise. Then in 1983, he had an epiphany: composers follow rules until they break them, and that gives their music life and energy. Cope developed an algorithm to insert some randomness into his rules-based music program. He also started expanding the factors the program analyzed and weighted to compose new music, including narrative tension, surprise, and storytelling. At this point, Cope began to feel as if the program, which he called Experiments in Musical Intelligence, had developed a personality.[64] In response, he began to call it Emmy. (As Emmy and Emily Howell demonstrate, he appreciates a good anthropomorphosis.)[65]

With enough input from a particular composer, Emmy could analyze and recombine it in new ways. Emmy became a prolific composer. One day, Cope left Emmy on while he went out for lunch; by the time Cope returned home, Emmy had written 5,000 original Bach chorales.[66] Cope believes that Emmy eventually produced thousands of works in the styles of 36 dead classical composers.[67] In 1993, Cope released an Emmy album, *Bach by Design*, which was followed by *Virtual Mozart* and *Virtual Rachmaninoff*, among others.[68]

In 2004, Cope retired Emmy. He had received too many calls from musicians who liked Emmy's compositions, considered playing them, but ultimately decided they weren't "special enough." Her prolific production rendered her products commonplace. Emily Howell is her replacement, or her "daughter," as Cope has called her. Howell has a huge library of music from which to draw influence and create rules. Its influences range from 16th-century Italian court music to Cope's compositions.[69] Its rules are the rules of that music, as Howell interprets them.[70]

Although Howell's output is more collaborative than Emmy's—Cope will ask Howell questions and answer yes or no to the musical statements it makes in response[71]—Cope maintains that Howell is its own musical master. "[Y]ou can't take it for walks. I can only generally pick it up and point it in the direction that I want it to go."[72] In this way, Cope is acting like a mentor to a promising student, tutoring Howell to help the AI reach its potential. Indeed, Cope believes that Howell is expressing itself in its music, producing musical ideas that he, as Howell's creator and programmer, never would.[73]

Howell's 2010 CD represents a triumph of this belief, as does the interest in Howell that commercially successful popular artists have expressed. A mainstream pop music group approached Cope around the time *From Darkness, Light* was released, looking into Howell's availability. (Cope will not reveal which group.) Cope believes that Howell will continue to evolve and develop a mature style. "Five years from now," he said when Howell's first CD was released, "I believe she will really be somewhere."[74]

Visual Art

As Cope used rules to program the creation of music into Howell, Harold Cohen used rules to program the creation of painting into Aaron, "the world's first cybernetic artist."[75] Cohen is a visual artist who, after studying painting at the Slade School of Fine Arts in London in the 1960s, moved to San Diego. There, he developed an interest in computer programming and artificial intelligence. The Stanford University Artificial Intelligence Laboratory, working on numerous AI projects at the time, including autonomous vehicles, invited him to work there as a Guest Scholar. His work with the AI Laboratory and in subsequent years focused on building a machine that could create the human act of drawing. That machine is Aaron.[76]

Cohen developed Aaron gradually over several decades. Aaron could first make crude "freehand, ad-hoc" drawings. Slowly, Cohen programmed specific figures into Aaron's program. In 1985, Aaron drew a representational figure of the Statue of Liberty that appeared in an exhibition on the history of the Statue. In the early 1990s, Cohen added color to Aaron's repertoire, permitting it to paint rather than merely sketch.[77] Now, Aaron is, in many ways, a fully functional artist. In addition to creating art on its own, it also mixes its own paints and washes its own brushes (while many of us struggle to clean our own dishes).[78]

Cohen has achieved this milestone through complicated and numerous rules governing and informing how Aaron paints, just as Howell and Emmy compose music based on rules Cope programs. Cohen insists this is how human artists create as well. "The vast majority of us," Cohen has said, "follow rules that somebody else taught us when we were growing up and going to art school ... The computer can in principle enact whatever rules you're capable of enunciating."[79] This is particularly important for Aaron because it can only rely on what it is told (i.e., what it is programmed) and cannot make use of associative knowledge. In other words, if Aaron is going to draw a kangaroo, Cohen must describe a kangaroo in explicit detail; he can't say "It's like a giant rat with a belly pouch."[80] But other than this input (and occasionally telling Aaron what *not to do* for expediency), Cohen does not give Aaron direction. "I don't tell it what to do. I tell it what it knows, and IT decides what to do."[81]

Cohen challenges critics who believe that Aaron does not make art. "If what AARON is making is not art, what is it exactly, and in what ways, other than its origin, does it differ from the 'real thing?'" he asks. "If it is not thinking, what exactly is it doing?"[82] Arguably, the fact that Aaron has prompted artistic criticism demonstrates its bona fides as an artist.

Regardless of any criticism, Aaron has found some creative and financial success as an artist. Its paintings have hung in major museums around the world, and consumers purchase Aaron's completed works.[83] "I do not believe that AARON constitutes an existence proof of the power of machines to think, or to be creative, or to be self-aware," Cohen has said. "It constitutes an existence proof of the power of machines to do some of the things we had assumed required thought, and which we still suppose would require thought, and creativity, and self-awareness, of a human being."[84]

Writing and Literature

In the first chapter, I made references to computer-generated poetry, novels cowritten by a reprogrammed Macintosh, and AI-written sporting reports. Although computers produced the content that was published in each example, only the sports reports represent actual AI writing. The others were prompted and heavily edited by humans. All, however, rely on rules to create.

In 1984, William Chamberlain published a book of poetry titled *The Policeman's Beard Is Half Constructed*, which he claimed was entirely written (outside of the introduction) by Racter (from raconteur), a program written in "compiled BASIC on a Z80 with 64k of RAM."[85] However, that is true only in the sense that Racter produced the words and phrases that became the book, but those words and phrases were prompted by templates that Chamberlain created.[86] Racter strung together words based on programmed structures and rules, but it was largely gibberish, meaning

Chamberlain had to perform significant editing for Racter's output to become poetry.[87]

Nearly a decade later, Scott French published *Just This Once*, a novel written in the style of Jacqueline Susann, author of *Valley of the Dolls* and a pioneer of glitz fiction. French spent eight years and $50,000 reprogramming a Mac to analyze two Susann novels and produce text based on her writing. He called the revamped machine "Hal," and attributed authorship of *Just This Once* to it, explaining that the book is "as told" to French by Hal.[88] In fact, French prompted and edited Hal just as much as Chamberlain prompted Racter, providing Hal with a basic plot, scenes, and characters. "I wrote several thousand rules to describe my characters," French explained at the time, "and then as the plot developed the computer told me what Ms. Susann would likely have written under the circumstances were the story hers."[89] Hal would ask French specific questions based on formulas derived from Susann's novels and use French's answers to produce a few sentences at a time. Then Hal would ask another question. "It doesn't write whole paragraphs at a time," he conceded at the time. "You can't get up, walk away, come back and find a completed chapter. It's not that advanced."[90] This contrasts with how David Cope has used Emmy, which can create whole pieces of music—thousands of pieces—autonomously.

More recently, innovative programmers have combined statistical analysis with writing programs to create AI that can draft many complete articles in a fraction of the time it would take a human author to complete one. Two companies in particular are leading this area of AI development, Narrative Science and Automated Insights. Both provide content (articles and analysis) for major sports properties. Narrative Science has worked with the Big Ten Network (a joint venture of the Big Ten Conference and Fox Networks) to produce short recaps of the conference's games—including baseball, softball, football, and basketball—within minutes of the conclusion of each game.[91] In 2012, Automated Insights entered into a deal with Yahoo Sports to provide more than 50 million automated fantasy football recaps, basically AP-style reports for all of the fantasy football games that Yahoo Sports facilitated. After each NFL week is over, Automated Insights drafts reports at a rate of about 300 per second for every Yahoo Sports fantasy football player. The reports read like they were written by a human reporter sent to cover each fantasy football matchup, providing stats from that weekend's NFL games, statistical history of each specific fantasy football league, and personalized information about the fantasy teams.[92]

Basically, these companies have developed sophisticated AI writing programs that search large amounts of data in order to identify trends, insights, and patterns. As the programs do that, they analyze the data and write reports in plain English that appear as if they were drafted by a human reporter.[93] Sports, with their numerous statistics and widespread popularity, are a good venue for this AI. It can analyze a massive set of data that would be impossible for a human being to analyze within the time constraints of a season while at the same time producing a near infinite amount of original material for the huge fan base to read. But there are other areas where these AI writers are also useful: financial news, corporate earnings previews, personal summaries of cell phone usage,[94] real estate listings, etc.[95] Narrative Science provides such content to publications like *Forbes* already.[96] Kristian Hammond, the chief technology officer of Narrative Science, believes that this AI technology could supply anywhere from 20–90 percent of a newspaper's content, including financial, sports, and real estate sections, as well as some entertainment stories based on box-office numbers.[97]

Obviously, AI writers currently thrive best when there is a lot of data to use as the basis for what they write. After analyzing the data, an AI writer chooses an article's tone based on its assigned perspective.[98] For example, Automated Insights operates fan pages for every NFL, MLB, NBA, and NCAA football and basketball team.[99] The AI writes as if it were a "homer" writer, sounding crestfallen if the team loses and triumphant if the team wins. Next, the program consults a huge database of phrases in search of words that match the story told by the numbers. If there is a lopsided score, the AI might write a headline like "Red Sox *Batter* Yankees." The few human writers that work for Automated Insights add phrases to this database for future use by the AI.[100]

Although I wondered how frequently mistakes occur in the finished narratives and reports, there are very few, if any. Hammond says, "People ask for examples of wonderful, humorous gaffes, and we don't have any." Similarly, Lewis Dvorkin, Forbes Media chief products officer, reports that there hasn't been a single reported error or complaint about the reports written by Automated Insights' AI writers, a better accuracy record than most of its human writers. Although the leaders of these companies do not expect human beings to be completely removed from the business of writing news, they do expect major advances in the sophistication of the finished written product.[101] "In five years," Hammond says, "a computer program will win a Pulitzer Prize—and I'll be damned if it's not our technology."[102]

At this point, AI writers are limited essentially limited to non-fiction and journalism. But the time may come when AI produces great works of fiction, fantasy epics, crime novels, and steamy romances. David Cope certainly think so. "I believe that without a doubt computer programs will write novels," he says. "Even great novels. It seems to me that we would be selling human creativity short if we didn't believe that to be true. For programmers are as creative as authors are."[103]

Who Owns IP Created by AI?

The AI described above demonstrates situations in which there is no human creator of a final work. David Cope mentored Emily Howell, but Howell created the music. Harold Cohen taught Aaron everything it knows, but Aaron created the paintings. Automated Insights and Narrative Science programmed a library of words and phrases into their AI writers, but the AI writers wrote stories and reports after conducting their own data analysis.

But as we discussed earlier in this chapter, copyrights (and patents) cannot be created by nonhumans. The words "author," "inventor," and "patentee" all refer to people, not machines or programs. What does that mean for the potentially valuable intellectual property rights created by AI? There are two basic options. We can either twist intellectual property law so that the human programmers receive the copyright or patent as if they themselves actually created it, or we can amend intellectual property laws so that they appropriately incorporate creative AI.

Intellectual Property Created by People Using AI-Created Parts

Admittedly, I'm putting a biased spin on the options. The people who think that the programmers of creative AI should receive full intellectual property law protection have their reasons. (And because using a straw man argument is using a losing argument, I will rely on William T. Ralston's "Copyright in Computer-Composed Music: Hal Meets Handel" article from a 2005 issue of the *Journal of the Copyright Society of the U.S.A.* as a well-articulated defense of this position for the remainder of this chapter.) Those reasons are not bad in some AI scenarios, such as Siri's involvement in the production of valuable copyrightable media, as discussed in chapter 1. However, the same reasoning makes their conclusions inapplicable to much of the technology above, in which a human merely turns the AI on, and 5,000 musical compositions appear over lunch or 300 sports stories write themselves in a second.

The thrust of Ralston's argument is that any human input that will affect the final creative product is sufficient for that product to be considered original under copyright law, which is necessary for the creation of an enforceable copyright.[104] Whoever provides the input into an AI creative process that produces "something unique" is, therefore, the author.[105] In coming to this conclusion, Ralston relies on several Supreme Court cases. In particular, two cases strengthen his point. In *Bleistein v. Donaldson Lithographing Co.*, the Court stated that "personality always contains something unique. It expresses its singularity even in handwriting, and a very modest grade of art has in it something irreducible, which is one man's alone. That something he may copyright."[106] In *Burrow-Giles Lithographic Co. v. Sarony*, the Court stated that "An author ... is he to whom anything owes its origin."[107] In other words, an author is the person who makes something unique, as its origin occurred when it became different from what it was before.

In analyzing these (and other related) cases, Ralston explains that there are two people with a claim of authorship: the programmer who creates the AI (for Siri, the Apple engineers) and the user who utilizes the AI to create new media (owners and users of individual iPhones with Siri). He concludes that we should rely on a "but for" rule to assign ownership of intellectual property to the person who uses—or even just turns on—an AI program that creates a new product or media. But for the user turning on the AI, Ralston argues, the new copyrightable material would never have been created. He argues that this supersedes the claim from the AI's programmer/developer to the copyright because the user of the AI can argue that he or she would have used another AI to create the copyrightable work.[108] So under his argument, David Cope would own the copyright to any of Emily Howell's or Emmy's works; Harold Cohen would own the copyright to any of Aaron's works; and Automated Insights and Narrative Science would own the copyright to the written works of their AI writers. David Cope agrees with this analysis of the music Emmy and Emily Howell produce. "I would argue that since I created the program that composes the music, choose (carefully) the music for its database, and choose works from the final outputs, that I have copyright responsibilities," he explains.[109]

The problem with this conclusion is that it ignores that the developers of these AI programs admit that the AI is the source of the original works and that they are not. David Cope claims that "The program is a cat not a dog. It keeps itself to itself, you can't take it for walks."[110] Harold Cohen admits that "I don't tell it what to do. I tell it what it knows, and IT decides what to do."[111] Ted Sullivan, whose software company GameChanger Media Inc. hired Narrative Science to generate game-recap stories, retained that company *because* there are no human creators. "We cover millions of games each year, so I would need an army of sports writers to write with the scale of the technology ... I don't know what it would cost to pay journalists to write that many articles, but I doubt we could afford it."[112] Under Ralston's analysis, a human author becomes the author or inventor of a new media or invention when he or she adds "something unique." But the people involved in the AI above admit that *they* are not adding anything unique; the programs themselves create "something unique."

Rather, Ralston's analysis is well-suited to the Siri-assisted media creation we discussed in chapter 1. If the owner of an iPhone experiments with Siri to obtain a particularly humorous or entertaining response that he or she uses to create a valuable piece of intellectual property—it is incorporated into a commercial jingle, it becomes a successful song, etc.—the user is entitled to the copyright. Although Siri's statement is part of the final product, it was the user who played with Siri and composed the final product that incorporated Siri's sound bite. Even though Apple engineers programmed the Siri software, the user actually authored the media by experimenting with Siri. Ralston believes that arguing that Apple is more responsible than the user for the valuable media because it created Siri is similar to arguing that a knife manufacturer is more responsible for a murder than the person who wielded the knife.[113]

I think he has something there. The real purpose of copyrights and patents under the Constitution is to "promote the Progress of Science and useful Arts."[114] If the initial programmer of AI is able to claim ownership of copyrights or patents that subsequent users of the AI create, those subsequent users will have little financial incentive to create. The progress of science and useful arts is promoted by providing financial rewards to users who use the AI well, not by hoarding any rewards with the developer.

Intellectual Property Created Solely by AI

The Siri scenario is entirely different than what Emily Howell, Aaron, and the AI writers do. Everyone involved with those products admits, one way or another, that the creative AI is creating the final product; no one claims that a human being is directing the AI. Even David Cope admits that Emily Howell is just expressing itself, even though he mentors it through the writing process.[115] When there is no direct human author, Ralston's argument falls apart because there is no human user to add "something unique." Although Ralston goes so far as to say that even turning the machine on makes the user the author, it is hard to see how rewarding the person who pushes the on-button "promote[s] the Progress of Science and useful Arts."

I think Ralph D. Clifford, whose analysis of AI-generated art was discussed in chapter 1, has the better argument. Clifford looks at intellectual property—both copyrights and patents—created by AI in essentially the same way that Emily Howell, Aaron, and the AI writers from Narrative Science and Automated Insights create new music, art, and content. He concludes that a claim of authorship by the user is unsustainable.[116] A human user of creative AI cannot claim a patent or copyright because "he or she did not conceive the invention" or media.[117] And, as explained above, AI cannot claim the patent or copyright because it can be neither the author nor the inventor under current law. Without someone to claim ownership of the intellectual property, it enters the public domain and anyone can use it.

But if AI creates something valuable—a hit song, a multimillion-dollar invention, a best-selling book that is optioned for a movie, etc.—*someone* will claim ownership. Someone will want the profits from that creation. The problem is that this will potentially make a mess of our intellectual property laws, as courts try to arbitrate claims from competing humans: users, programmers, and other people who want to profit from the AI's capability.

Addressing AI

The current intellectual property laws cause confusion regarding creative AI because they do not address AI. People will try to take advantage of that confusion for profit, which will stunt the potential development of AI and new intellectual property developed by AI. We should revise both public and private laws to fix this problem.

Changes to Intellectual Property Laws

The consequences of who owns AI-created intellectual property are potentially far reaching, shaping incentives for creating and using AI and directing the wealth generated by AI. Inconsistent court decisions will only exacerbate this problem. Congress should amend existing intellectual property laws to address creative AI by expanding the definition of author and inventor to accommodate music, art, content, etc., and inventions that are created by AI. I would suggest a simple two-part system:

1. If the AI depends on human interaction to create, the user is the author or inventor as if it had created the intellectual property without the AI;
2. If a person can turn on the AI and it creates without further human direction, the AI is the author or inventor, but the creator of the AI

owns the copyright or patent for 10 years before it enters into the public domain.

The first part encourages people to continue playing with Siri and similar devices by recognizing that users contribute something unique to the final product and rewarding them with the traditional copyright protection afforded to creators who are not assisted by AI.

At the same time, the second part encourages programmers to continue developing AI that can constantly and autonomously create new content and media, while also enriching the public at large by placing those creations in the public domain much faster than works created solely by a human being. A programmer who develops an AI writer that can create a new *Harry Potter*, *Twilight*, or *Hunger Games* franchise every day would get ridiculously rich, as those franchises would launch movies, television shows, toys, posters, lunch boxes, etc. If the programmer received traditional copyright protection for her machine's creations, she and her family would be fabulously rich for generations, even though no person wrote the stories that made them rich. Retaining the traditional protection does little to "Promote the Progress of Science and useful Arts." If the AI's works pass into the public domain after 10 years—granting an appropriate period for the programmer to make substantial money from the works of her creation—then everyone will be free to use the AI's works without any restrictions after a decade. It would be the liberation of fan fiction, and possibly its monetization too. With a financial incentive, people will be more likely to use the AI's creations in further creative works, achieving the constitutional goal of our intellectual property system.

In general, this two-rule system would assign the ownership of intellectual property created by or with AI as I've discussed above. The lone exception: David Cope's Emily Howell. Cope's Emmy could compose music completely autonomously, but although Howell writes music, it requires someone to interact with while composing.[118] However, Cope's interactions with Howell are extremely time consuming[119] and are getting longer as their musical relationship develops,[120] addressing one of my primary concerns about creative AI: the production of new intellectual property in a fraction of the time needed for human creativity.

Policies for Companies That Develop Creative AI

In light of the uncertain status of current intellectual property laws, companies like Narrative Science and Automated Insights should draft company policies providing an outline of the AI they use (although nothing that would disclose trade secrets or proprietary information) and a clear analysis of how intellectual property law applies to their content. Until Congress updates our IP laws, the analysis will provide greater protection to these companies from potential threats to their copyrights and patents. It can state the relationship the companies expect between themselves, their programmers/employees, and their users and also how they believe IP laws govern the content they produce and their ownership of that content. Expectations are important, so every employment contract and user agreement should include that analysis, providing notice to every party that interacts with the content and protecting the companies in any potential lawsuit.

Even after Congress passes new intellectual property laws that incorporate AI, companies that develop creative AI should retain those analyses, updated to address the new laws.

The Work of Human Hands versus Pumping It Out Like a Spigot

I recommend cutting the lifetime of copyrights and patents short because the traditional incentive for a longer life span is not present. A musician can work for years on a single album; an inventor can similarly spend much of his or her life on one breakthrough patent. Creative AI can do that same work in a fraction of the time. As David Cope put it, Emmy could "pump out music like a spigot."[121] There is no need to provide the same reward we have reserved for people who dedicate years of their lives to creativity and research to people who make machines that can condense their years into a morning, an hour, a minute, or less. Owning a copyright or patent for multiple decades is important when you have invested years of your life in that one piece of intellectual property. It is less important when it is one of hundreds or thousands of potential pieces of intellectual property created over breakfast.

At the same time, we want to encourage programmers and developers who master and advance creative AI. The two rules above create an easy-to-administer system that will promote the progress of science and the useful arts by fairly rewarding people who work with or develop creative AI. Developers of AI like Siri will have the incentive to continue developing similar AI devices in order to sell them to users. Users of AI like Siri will have the financial incentive to create using the AI because they can potentially profit from the resulting intellectual property. Developers of more sophisticated creative AI like those discussed in this chapter will be rewarded for their machines and programs, but the public at large will also benefit because their works will enter into the public domain much sooner than traditional intellectual property.

The goal in this system is to ensure that we promote the development of AI by rewarding talented programmers while also trying to spread the benefits of AI to as many people as possible. The next chapter explores this concept more generally as it summarizes the ideas that this book has presented.

CHAPTER 9

Can AI Be Good for Us?

Occasionally, a friend or family member will object to the idea of an autonomous car because people will ride in cars and forget how to drive, which they believe represents a form of unhealthy dependence. On the other hand, those same friends and family buy cheese at Whole Foods and can't milk a cow, but don't consider themselves helpless. Technological advances—whether AI in cars or cheese at the store—have the ability to make our lives better. Some personal independence is part of the cost, but the return can be greater time for human beings to reach our potential. In 1850, about 64 percent of Americans worked on a farm;[1] by 2008, fewer than 2 percent of Americans worked on a farm.[2] Technological changes permitted millions of Americans to leave the fields and become the writers, doctors, engineers, and entrepreneurs who have made modern life possible. AI is what's next.

Each of the preceding chapters discussed a form of AI that either is already functioning in the world or is likely to be available in the near future. A lot of this is revolutionary technology that will drastically change some aspects of modern life. Innovations like driverless cars and "robonannies" potentially create loads of free time for human beings. AI in cars and robotic surgical systems hold out the promise of saving millions of lives by removing human error. AI industrial workers could make manufacturing affordable to more entrepreneurs by offering labor at a fraction of the current cost. In the not-so-distant future, AI drones will be able to monitor us—in the United States and abroad—while we enjoy articles written by AI without any human input.

I sometimes have to remind myself that these developments aren't good or bad in and of themselves. We get to determine whether they're good or bad by how we respond. And that doesn't just mean how we individually use AI, whether we drive ourselves or let the car drive us. We collectively get to decide how the benefits of AI will be shared through public policy decisions, regulations, and laws. In response to the Industrial Revolution of the 19th century, we spent much of the 20th century ensuring that a large and prosperous middle class emerged by changing our laws to require a minimum wage, limit work hours without overtime, prohibit child labor, and protect environmental health and workplace safety.

We need to do that now to ensure that AI helps to create a rejuvenated 21st-century middle class in much the same way the laws addressing the Industrial Revolution created the 20th-century middle class. However, given the speed of technological advances, we'll need to respond to AI much faster. To illustrate the difference between effective laws and ineffective laws governing AI, the next sections will look at the best- and worst-case scenarios (using details drawn from reports, analyses, and possibly some fictional accounts) before outlining public policy goals and changes to our laws that can lead to the best-case scenario.

Worst-Case Scenario

Historically, there is a correlation between technological advances and economic growth. Advances in agriculture permitted greater food production. Advances in manufacturing expanded the supply of fabricated goods like furniture and clothes. However, the relationship between technology development and increased income is not a rule. Economists have looked at recent and pending advances, like the Internet and AI, and questioned whether the correlation will continue. "We have a collective historical memory that technological progress brings a big and predictable stream of revenue growth across most of the economy," writes George Mason University economist Tyler Cowen in *The Great Stagnation*. "When it comes to the web, those assumptions are turning out to be wrong or misleading."[3]

It's a legitimate concern that the same will be true of AI. "There's no economic law that says everyone has to benefit equally from increased productivity," writes M.I.T. professor Erik Brynjolfsson. "It's entirely possible that some people benefit a lot more than others or that some people are made worse off."[4] With this idea in mind, what follows are possible worst-case scenarios that AI could lead to:

- Fifty million professional jobs are lost to AI[5] (compared to only 7.5 million jobs lost during the Great Recession),[6] representing

40 percent of the workforce,[7] including lawyers, doctors, writers, and scientists;[8]

- New jobs aren't created to replace those jobs;[9]
- The wealth from AI is accumulated in the hands of the few people who own and design AI programs and robots, resulting in even sharper economic inequality than currently exists;[10]
- Human interaction is greatly devalued—and fewer people notice you're gone—because AI will keep sending messages for you after you die;[11]
- Malevolent AI will come to dominate the world as it grows smarter than people, threatening the lives of our grandchildren;[12]
- AI engages in open combat with mankind, killing the vast majority of humanity and forcing a rebel human resistance to wage guerrilla warfare in a near-hopeless effort to save the species;[13] and
- Mankind is trapped by sophisticated AI robots in a prison that taps directly into the human mind, so that the machines can live off the electrical activity in our brains.[14]

Admittedly, the last few examples are given tongue-in-cheek. But the first few are potentially real and devastating. Job losses more than six times worse than the Great Recession, with no relief in site because AI performs the potential replacement jobs too. Wealth concentrated solely in the hands of the few people who create and own AI programs and machines, with little trickling down to everyone else because the new technology does not create new jobs to benefit and grow the middle class. And there might be less actual human warmth in the wake of widespread AI use as our interactions are increasingly with AI, even to the point where AI steps in to communicate for us after death. There will be no need to appreciate our loved ones; they will always be there to tweet to us.

Best-Case Scenario

But AI has the potential to do much more good than harm, if we use it in the right way. The worst-case scenario depends on not doing anything in response to the development of AI: no public policy changes, no amended laws, no new regulations. The worst-case scenario also ignores the almost certain benefits of many types of AI, like autonomous vehicles.

Brynjolfsson from M.I.T. questions the historical relationship between technological advances and widespread prosperity, noting that there is no "economic law" that mandates that as technology advances, everyone benefits. And he's right. It hasn't been an economic law forcing technological developments to broadly improve our quality of life. Man-made laws have done that.

As we've discussed, the Industrial Revolution created millions of jobs, but also hurt the people who worked in factories through dangerous working conditions, environmental hazards, and child labor. At the same time, the owners of the factories became extremely wealthy, despite the dangers they introduced to their workers and the environmental damage caused by their factories. The full benefits of the era's technological advances were not experienced by the majority of workers and Americans until laws were introduced to mandate safe workplaces, to protect the environment, to prohibit child labor, etc. Similarly, other technological advances introduced benefits that were diminished by their dangers, which were eventually addressed by subsequent legislation. For example, automobiles permit us to travel much more freely than we did before cars, but they also kill tens of thousands of people each year.[15] However, the number of vehicle-related deaths has been significantly decreased by laws requiring seat belts.[16]

Assuming we make the necessary adjustments to our laws, here are some best-case scenarios for a world with the widespread adoption of AI:

- As current automobiles use only 8 percent of the road at most at any one time, autonomous vehicles will at least double the percentage of the road used by cars because they can drive more closely together than human drivers, eliminating traffic jams completely;[17]
- Autonomous cars reduce the number of car accidents by 90 percent;[18]
- The military's use of AI drones with advanced surveillance and targeting capabilities leads to a net decrease in casualties among soldiers and civilians in war;[19]
- AI doctors provide medical assistance to people and places where human doctors are not always sufficiently available, like accident scenes, flights, war zones, and developing countries;[20]
- Cities are redesigned to integrate AI, making our lives easier, more enjoyable, clean, green, and safe because there is less reliance on human whim and error;[21]
- AI relieves a large percentage of human beings of the need for regular employment, providing material goods automatically and giving them greater freedom to read and engage in education and self-improvement;[22] and
- Without a need to maintain jobs, human beings focus on each other, and "Friendship could become a living art again."[23]

Obviously, some aspects of the best-case scenario look almost as unlikely as parts of the worst-case scenario, and the idea that AI turns everyone's life into *The Jetsons* is a false promise. But the potential benefits suggested above are real: AI can improve the quality of our lives, giving us more time for friends, family, and reaching our human potential. This is essentially what happened after the Industrial Revolution, as its technological developments freed millions of people to attend school, creating the engineers and scientists responsible for the modern era. In order to leverage AI appropriately, we have to revise our laws and public policy to reflect a legal personhood for AI programs and robots.

Public Policy Changes

Before discussing the changes to our laws to address AI, let's say a few words about the public policy behind those changes. Although laws need to be changed to treat AI as a legal person in certain situations, those changes should only be made to encourage the development of AI in such a way that its benefits are spread widely to strengthen and grow the middle class. The development of AI should not be the goal in and of itself.

This means providing incentives to the developers of AI, protecting the consumers who use AI or have AI used on them, regulating how AI is used by consumers and professionals, and encouraging professionals to adopt AI. So a law establishing mandatory AI reserve funds that are financed by a share of each AI purchase is not desirable because it prevents consumers from getting screwed. It is desirable because it both protects consumers (by giving them guaranteed money in case of a problem with the AI) while also providing an incentive to developers, who will know the degree to which their liability exposure is limited. Although the end result is the same, the reasoning matters. It will help legislatures and regulatory agencies create appropriate and genuinely helpful laws and regulations.

Similarly, AI should be treated as a legal person at times because that treatment furthers the goals listed above or helps enforce the constitutional use of the technology. When a court considers police officers' use of an AI surveillance drone under the Fourth Amendment, the judge should look at whether the drone's activities would be constitutional had a human performed them. Not because this gives the police wider freedom to spy—it won't always—but because this helps courts remember that the reasonable privacy the Fourth Amendment protects is an important human right. Technology that acts like a human should be held to at least the same human standard of privacy.

In trying to strengthen and grow the middle class, we have to remember the effect of AI on jobs. AI, by design, will destroy jobs. This is not necessarily bad, but as a society we have to have a plan for addressing the large number of people that will either lose a job or will come of age and be unable to find one. Although I'm sure there will be many ideas to help displaced workers, I think we should consider fundamentally reexamining how we think of people who work. Our education system—from public school through college, adult learning, and training programs—is geared toward creating employees. Our students learn skills that employers want; laid-off workers are encourage to receive training to make themselves more employable. If AI destroys jobs by design, we need to revise education to stop creating employees (people who seek jobs from others) and begin creating entrepreneurs (people who create jobs for themselves). This will require introducing new, core courses about entrepreneurship and business formation into our public schools and colleges. It will require a collective reassessment of what risk adults find acceptable in their working lives. And it will require major government action, as well as cooperation from the existing businesses that will gain by adopting AI. However, after we change our laws to appropriately accommodate AI, I believe an emphasis on entrepreneurship and small business start-ups is the best public policy to ensure that the middle class benefits from an economic and legal landscape where AI is widely used and welcomed.

How We Should Change Our Laws to Address the Emergence of Artificial Intelligence

Just like the technological advances of the Industrial Revolution, AI is coming. We cannot stop it, and if we respond appropriately, we shouldn't want to. Instead, we should try to make sure that we properly guide its adoption. You didn't have to read the last eight chapters too carefully to come away with some specific ideas about changes to our laws that will accommodate AI and spread its benefits widely. The basic idea running through these legal changes is that because AI will make decisions like a human being—and because we will at times interact with AI as a human being—it is appropriate to treat AI like a legal person with legal rights and obligations in certain situations. By doing that, we'll make sure that AI's benefits are experienced by everyone. To quickly summarize legal changes—to public and private law—proposed in the book, here's an outline:

Liability Involving AI

- For certain forms of AI that physically interact with the world (cars, construction workers, surgeons, etc.), liability should be assigned solely to the AI when no other party is at fault. To permit the AI to pay when it is liable, states should require that either 1) insurance covers each relevant AI or 2) there be a reserve fund that is funded by the collection of a small fee when each AI is purchased and that pays for damages caused by relevant AIs.

- In order to avoid costly lawsuits and very high standards of proof that may unreasonably prevent victims from recovering for damages caused by AI, states should consider implementing a payment system for liable AI similar to the worker's compensation system. The standard of evidence necessary to be paid on behalf of the AI would be lower: victims only need to show actual injury or loss and reasonable proof that the AI caused the injury or loss. But in return for easier and faster payments, the payments would be lower than what might be possible in court. This permits the victims to recover faster and easier while also letting AI developers and manufacturers plan for an established potential loss.

- Companies that use and manufacture AI should establish written policies governing how the AI should be used, who is qualified to use it, and what operators and other people can expect from the AI. This will help to give the human operators and beneficiaries an accurate idea of what to expect from the AI while also protecting the companies that make the AI from future litigation.

- States should not automatically assign liability to the person who turns on the AI. If it is appropriate to assign liability to a person involved in the AI's operation, it is most likely the person who oversees or manages the AI while it operates, who is not necessarily the person who turned it on.

Regulation of AI

- Human oversight of AI should only be required when the primary purpose of the AI is to improve human performance or eliminate human error. When the primary purpose of the AI is to provide for human convenience, like autonomous cars, requiring oversight defeats the purpose of the AI.

- AI should be required to have an easy-to-use off switch. Similarly, AI products that perform physical functions should have a warning system to let nearby people know that the AI is malfunctioning and that they should use the off switch. The AI should be able to turn itself off if no human is around to stop it while it malfunctions.
- States should require license endorsements to use AI only if specialized knowledge of the AI's operation is necessary (e.g., surgery, plane piloting, etc.) and human lives are particularly vulnerable. License endorsements for other forms of AI are likely unnecessary, especially if a human would otherwise need a license to use the device, like a car.
- Manufacturers and developers of AI must disclose to purchasers what information the AI collects during its use. States may want to consider requiring uses agreements in which the purchaser can opt out of any data collection but receives a user fee from the manufacturers for any data collected.

Powers of Attorney

- Anyone who has a durable power of attorney may want to revisit the terms of that document to ensure that they accurately reflect his or her feelings about AI caregivers. If the idea of a robot assisting you later in life or with the care of your children bothers you, you should make sure your power of attorney says so.

Municipal and Zoning Ordinances

- To compensate for lost parking revenues when autonomous cars stop parking in public parking areas, cities and towns should consider a congestion toll to enter downtown or business areas, which would increase or decrease depending on the time of day. Workdays would have a higher toll; weekends would have a lower toll. The toll can be collected automatically with an EZ-Pass, and residents who actually live in the areas with the new tolls should be exempt from paying.
- If towns and cities are worried about their rural character or scenic vistas, they should revise their zoning ordinances to carefully govern what forms of AI can be used on farms and in rural zones.
- Zoning ordinances should be revised to permit manufacturing in commercial or downtown areas when the AI is appropriate to the area. Quiet AI that does not create noise, odors, excessive parking

or traffic should be acceptable in those areas and can help to maintain or increase economic activity in a downtown area.

Fourth Amendment

- When considering whether surveillance drones used by police or other government agencies are constitutional, courts should consider whether the actions of the AI drone would be constitutional had a human police officer done them.
- Companies that develop and manufacture AI surveillance drones should draft analyses of the constitutionality of their products' uses. Similarly, police departments that incorporate AI surveillance drones into their practices should prepare a similar analysis of the drones' constitutionality. This will help the court in any litigation that alleges the drones have violated the Fourth Amendment by preemptively explaining why the parties responsible for the drone believe its use is constitutional.

International Law

- The international community should organize a treaty that affirms nations are liable for the actions of their AI in the same way they are liable for the actions of their military personnel.
- A treaty should affirmatively permit AI cars, boats, and planes. The same treaty should establish standards for the maintenance of those vehicles, as well as for their drivers.
- Nations should draft multilateral agreements that address how AI impacts national sovereignty, what surveillance AI drones are permitted to perform, and what AI is permitted to do during armed conflicts.

Intellectual Property

- When an AI product relies on human interaction to create new content or inventions, the human user is the author or inventor and receives the same intellectual property protection as if he or she had created the content or inventions without any help from AI.
- When AI can produce new intellectual property autonomously without any meaningful human interaction (beyond the on/off button), the AI is the author or inventor, but the AI's creator will own the copyright or patent for 10 years before it enters into the public domain.

- Developers and manufacturers of creative AI should prepare clear analyses stating how intellectual property law applies to the intellectual property created by the AI. Similarly, every employment contract and user agreement should incorporate this analysis so that there is no confusion about the position the developers and manufacturers take regarding the AI's intellectual property. This will help them in court if there is litigation involving AI-created intellectual property.

Please note that these suggested changes are not exclusive. There are areas of law that this book does not address or only addresses in passing. The former includes the use of AI to analyze datasets to assist police in determining who to arrest and when. The latter includes surveillance drones used by individuals, not the government, and revisions to state privacy laws to govern those drones. Although the FAA's upcoming regulations will supersede any state or local attempts to govern the flight of AI drones, states may be able to limit what equipment can be installed and how the drones may be used. State lawmakers should consider bills addressing those topics.

Some of these legislative and regulatory changes would be difficult and time-consuming in an ideal environment. It might be damn near impossible right now and for the immediate future. Even fairly reasonable ideas that should avoid political controversy have not been able to in Washington, D.C. Pass a budget to keep the government running? No. Reform an immigration system that no one is satisfied with? No. Try new ideas to improve public education? No. Address AI quickly enough to ensure that the middle class benefits from its development? Probably not. Address AI nimbly enough to adjust laws appropriately as the technology develops? Absolutely not.

It took 100 years to enact laws that widely spread the benefits of the Industrial Revolution to a prosperous middle class. To repeat a major point of this book: We don't get 100 years any more. We have 20–30 years, tops, before the next big technological advance comes along. If we don't sufficiently address AI when we can, its benefits might never fully enrich the middle class. At the same time, I'm cognizant of the danger of legislating new technology sooner than appropriate.[24] I believe a reasonable compromise between doing too little too slowly and doing too much too quickly is achievable through an amendment to the Administrative Procedures Act.

Congress passed the Administrative Procedures Act (APA) in 1946. Among its provisions, it generally requires that federal agencies publish proposed rules in the Federal Register,[25] hold hearings to consider rules,[26] and permit comments and responses before any proposed rule can become effective.[27] Although the strict provisions of the act do not sound very demanding, in reality they become very time and energy consuming, potentially slowing the development of regulations governing AI. For example, when the Food and Drug Administration (FDA) looked into establishing a rule governing the percentage of peanut butter that is actually composed of peanuts—87.5 percent versus 90 percent—it generated 7,700 pages of comments for the FDA to review.[28]

My point is not that the APA is a waste of time. The APA ensures that regulations are well considered and publicized before they become effective. In general, that's a good thing. But when quick and responsive action is needed, the APA can be counterproductive. In order to expedite regulatory changes to address AI, it is worth considering an amendment to the APA that would permit a federal agency—maybe the Consumer Protection Agency—to quickly revise regulations affecting AI. As the technology develops, this would also permit further revisions to make sure that the existing regulations do not become outdated and detrimental to AI's development.

The APA would still publish in the Federal Register notice of rules affecting AI, but the notices would state that new rules are effective, not proposed. Although this would not provide the same safeguards against regulatory overreach, inappropriate rules will receive publicity and can be overturned just as quickly as passed.

Obviously, many of the legislative changes proposed above require action by Congress, but not all. An agency with the ability to promulgate and revise regulations quickly can respond to the needs of the middle class as AI appears in more products and its use becomes more widespread.

"No Need Either of Apprentices for the Master Workers or of Slaves for the Lords"

It's easy to oversell the promise of AI. Our science fiction stories have done it for decades. The Asimovs, Roddenberrys, and Dicks of the world promised us robots that will do everything for us that we don't want to do, from manual labor to interpersonal relationships we'd rather avoid. And after all this time, all we have to show for it is oversized expectations. But the robots, programs, and other AI discussed in this book show that this time, the potential may actually be realized because so much research and development have occurred to get us to the cusp of widespread AI. Not C-3PO or the *Bicentennial Man*, but AI that is much more modest.

We won't see self-aware robots that can chauffeur us around town, but the car can do that on its own.

Ironically, the robots aren't just coming for the jobs we don't want to do but for the jobs we to do want as well. AI writers, lawyers, doctors, etc. are coming, potentially displacing a large number of workers in professions that aren't used to getting displaced. Collectively finding new ways for those people to meaningfully spend their time may end up being the greatest challenge AI creates, as weak AI could potentially provide the labor necessary to feed, clothe, shelter, and entertain us. We'll have everything in our lives but meaning.

However, I believe shifting our educational goal from the creation of employees to the creation of entrepreneurs is an effective way to prevent that from happening while also allowing AI to benefit everyone, from rich and poor, urban and rural. I am also comforted by the opinion of Margaret Boden, who believes that AI can be "rehumanizing" rather than "dehumanizing."[29] She writes:

> [AI's] contributions to our food, shelter, and manufactured goods, and to the running of our administrative bureaucracies can free us not only from drudgery but for humanity. It will lead to an increased number of "service" jobs—in the caring professions, education, craft, sport, and entertainment. Such jobs are human rather than inhuman, giving satisfaction not only to those for whom the service is provided, but also to those who provide it. And because even these jobs will very likely not be full-time, people both in and out of work will have time to devote to each other which today they do not enjoy.[30]

That's the ideal. Revising our laws so that robots are people too is not just a way to make the middle class richer and stronger, although that can happen if we amend our laws properly. If we treat robots like humans, they'll let us be more human.

Notes

Chapter 1

1. Roger Clarke, "Asimov's Laws of Robotics: Implications for Technology Part 1," *IEEE Computer* (December 1993): 53–61; Clarke, "Asimov's Laws of Robotics: Implications for Information Technology Part 2," *IEEE Computer* (January 1994), 57–66.

2. John Chipman Gray, *The Nature and Sources of the Law* (New York: Columbia University Press, 1909), 27.

3. Portions of this section have been adapted from the author's article "Siri Is My Client: A First Look at Artificial Intelligence and Legal Issues," *New Hampshire Bar Journal* 52:4 (Winter 2012): 6–10.

4. "Apple Siri FAQ," Apple, http://www.apple.com/iphone/features/siri-faq .html.

5. Although there are GPS devices that exhibit advanced weak AI, reviews are mixed at best. By all measures, Siri's voice interface seems superior to that of any GPS device. On top of that, Siri's potential for ubiquity distinguishes it from GPS, which has a more limited function.

6. For the purposes of this book, texting, instant messaging, e-mailing, etc., are not considered man's "natural setting."

7. Steve Hehn, "Speak Up! Advertisers Want You to Talk with New Apps," *NPR*, April 15, 2013, http://www.npr.org/blogs/alltechconsidered/2013/04/15/ 177345718/speak-up-advertisers-want-you-to-talk-with-new-apps.

8. Farhad Manjoo, "Siri Is a Gimmick and a Tease," *Slate*, November 15, 2012, http://www.slate.com/articles/technology/technology/2012/11/siri_vs _google_the_search_company_s_voice_recognition_program_gets_closer.html.

9. John Markoff, "A Software Secretary That Takes Charge," *New York Times*, December 13, 2008, http://www.nytimes.com/2008/12/14/business/ 14stream.html?_r=1.

10. Timothy Hay, "Apple Moves Deeper into Voice-Activated Search with Siri Buy," *Wall Street Journal Blog*, April 28, 2010, http://blogs.wsj.com/venturecapital/2010/04/28/apple-moves-deeper-into-voice-activated-search-with-siri-buy/.

11. "Siri Launches Virtual Personal Assistant for iPhone 3GS," SRI International, press release. February 8, 2010, http://www.sri.com/news/releases/020510.html.

12. Hay, "Apple Moves Deeper into Voice-Activated Search with Siri Buy."

13. Jill Duffy, "What Is Siri?" *PC*, October 17, 2011, http://www.pcmag.com/article2/0,2817,2394787,00.asp.

14. Andrew Nusca, "Say Command: How Speech Recognition Will Change the World," *Smart Planet*, November 2, 2011, http://www.smartplanet.com/blog/smart-takes/say-command-how-speech-recognition-will-change-the-world/19895?tag=content;siu-container.

15. Duffy, "What Is Siri?"

16. *Bridgeport Music, Inc. v. Dimension Films*, 410 F.3d 792, 798-99 (6 Cir. 2005).

17. *Campbell v. Acuff-Rose Music*, 510 U.S. 569 (1994) (ruling that 2 Live Crew's use of Orbison's "Pretty Woman" constituted fair use).

18. *Newton v. Diamond*, 349 F.3d 591 (9th Cir. 2003) (ruling that the Beastie Boys were not liable for sampling James Newton's "Choir" in their track "Pass the Mic").

19. *Bridgeport Music*, 410 F.3d at 798–99 (ruling that NWA's sampling of a guitar chord from Clinton's "Get Off Your Ass and Jam" violated the copyright on that song).

20. *Bridgeport Music*, 410 F.3d at 795.

21. Although 17 USC § 106 distinguishes between the rights held by the owner of a musical composition copyright and those held by the owner of a sound recording copyright, that distinction is not relevant to this article.

22. 17 USC § 201(a), available at http://www.gpo.gov (accessed April 29, 2013).

23. *Community for Creative Non-Violence v. Reid*, 490 U.S. 730, 737 (1989).

24. See Melville B. Nimmer and David Nimmer, *Nimmer on Copyright*, vol. 1 (Newark, NJ: Matthew Bender, 2010), § 5.01[A].

25. Racter, *The Policeman's Beard Is Half Constructed* (New York: Warner Software/Warner Books, 1984).

26. Scott French and Hal, *Just This Once* (New York: Random House Value Publishing, 1993).

27. The program referred to is "Emily Howell," written by a UC Santa Cruz music professor named David Cope. Cope and Emily are discussed in greater detail later in this section.

28. Clifford is a professor at the University of Massachusetts School of Law. Nimmer and Nimmer rely on him as an expert in the evolving discussion regarding intellectual property produced by AI.

29. Ralph D. Clifford, "Intellectual Property in the Era of the Creative Computer Program: Will the True Creator Please Stand Up?" *Tulane Law Review* 71 (June 1997): 1675–1703.

30. Clifford, 1686–94.

31. Clifford, 1694–95.

32. *Feist Publications, Inc. v. Rural Telephone Service, Co.*, 499 US 340, 345 (1991).

33. Clifford, 1694, omitting citations.

34. Clifford, 1695. Clifford dismisses the idea that the AI itself could own the copyright, at least under the current law. Noting that the "author" of a work owns the copyright under the federal Copyright Act, he reviews its use there and in other portions of the U.S. Code. He concludes that "the use of the term 'author' in the Copyright Act implies Congress meant a human author ... the general use of the term 'author' in the U.S. Code reinforces the conclusion that Congress intended the term to mean humans." Clifford, 1682, 1684.

35. Clifford, 1695.

36. Jacqui Cheng, "Virtual Composer Makes Beautiful Music—and Stirs Controversy," *ars technica*, September 29, 2009, http://arstechnica.com/science/news/2009/09/virtual-composer-makes-beautiful-musicand-stirs-controversy.ars.

37. Ibid.

38. Ibid.

39. Farhad Manjoo, "Will Robots Steal Your Job?" *Slate*, September 27, 2011, http://www.slate.com/articles/technology/robot_invasion/2011/09/will _robots_steal_your_job_4.single.html.

40. If that seems too short a period of time, consider this: *Forbes* magazine estimated J. K. Rowling's net worth to be approximately $1 billion seven years after *Harry Potter* first appeared on bookshelves.

41. Chris Brooke, "'I Was Only Following Satnav Orders' Is No Defence: Driver Who Ended Up Teetering on Cliff Edge Convicted of Careless Driving," *Daily Mail,* September 16, 2009, http://www.dailymail.co.uk/news/article-12 13891/Driver-ended-teetering-cliff-edge-guilty-blindly-following-sat-nav-directions .html.

42. "Sat Nav Confusion Blamed for Penzance Crash," *Falmouth Packet.* Jul 3, 2012, http://www.falmouthpacket.co.uk/news/9793476.Sat_nav _confusion_blamed_for_Penzance_crash/.

43. "Driver 'Had Too Much Faith in Satnav' in Run Up to Death of Biker in Appalling Weather, Says Judge," *Daily Mail*, December 31, 2012, http://www .dailymail.co.uk/news/article-2080456/Judge-blames-drivers-reliance-sat-nav -jails-killing-motorcyclist.html.

44. Eric Sinrod, "Is GPS Liability Next?" *CNET*, January 16, 2008, http:// news.cnet.com/Is-GPS-liability-next/2010-1033_3-6226346.html?tag=ne.fd.mnbc.

45. Martin J. Saulen, "'The Machine Knows!' What Legal Implications Arise for GPS Device Manufacturers When Drivers Following Their GPS Device

Instructions Cause an Accident?" *New England Law Review* 44:1 (Fall 2009): 189.

46. Tom Vanderbilt, "Let the Robot Drive," *Wired*, January 12, 2012, http://www.wired.com/magazine/2012/01/ff_autonomouscars/all/1.

47. Vanderbilt, "Let the Robot Drive."

Chapter 2

1. See Dan B. Dobbs and Paul T. Hayden, *Torts and Compensation: Personal Accountability and Social Responsibility for Injury*, 5th ed. (St. Paul, MN: West Publishing, 2005), 2.

2. F. Patrick Hubbard, "Regulation of and Liability for Risks of Physical Injury from 'Sophisticated Robots,'" http://robots.law.miami.edu/wp-content/uploads/2012/01/Hubbard_Sophisticated-Robots-Draft-1.pdf (paper presented as a work-in-progress at We Robot Conference, University of Miami School of Law, April 21–22, 2012): 8–9.

3. "Rob" is unlikely to appear in court, anyway.

4. Dobbs and Hayden, 49.

5. Ibid., 148.

6. See American Law Institute. *Restatement (Second) of the Law of Contracts* (St. Paul, MN: American Law Institute Publishers, 1981), § 34 and Comments.

7. Dobbs and Hayden, 695.

8. Hubbard, 16–17; see David G. Owen, *Products Liability Law*, 2nd ed. (St. Paul, MN: West Publishing, 2008), 527–29.

9. Hubbard, 17; see *Williamson v. Mazda Motor of America, Inc.*, 131 S. Ct. 1131 (2011).

10. American Law Institute, Restatement of the Law, Torts—Products Liability (St. Paul, MN: American Law Institute Publishers, 1998), §2.

11. Ibid., §2(a).

12. See Dobbs and Hayden, 382.

13. *Verdicchio v. Ricca*, 179 N.J. 1, 23 (2004).

14. Dobbs and Hayden, 624.

15. "Look, No Hands," *The Economist*, September 1, 2012, http://www.economist.com/node/21560989.

16. "New Allstate Survey Shows Americans Think They Are Great Drivers—Habits Tell a Different Story," Allstate Insurance Company, press release, August 2, 2011, http://www.allstatenewsroom.com/channels/News-Releases/releases/new-allstate-survey-shows-americans-think-they-are-great-drivers-habits-tell-a-different-story.

17. Jim Motavalli, "Self-Driving Cars Will Take Over by 2040," *Forbes*, September 25, 2012, http://www.forbes.com/sites/eco-nomics/2012/09/25/self-driving-cars-will-take-over-by-2040/.

18. See Marina Koren, "How Raven, the Open-Source Surgical Robot, Could Change Medicine," *Popular Mechanics*, February 28, 2012, http://www

.popularmechanics.com/science/health/med-tech/how-raven-the-smart-robotic
-helper-is-changing-surgery.

19. Anne Eisenberg, "When Robotic Surgery Leaves Just a Scratch," *New York Times,* November 17, 2012, http://www.nytimes.com/2012/11/18/business/single-incision-surgery-via-new-robotic-systems.html?_r=0.

20. Royal Academy of Engineering, *Autonomous Systems: Social, Legal and Ethical Issues,* http://www.raeng.org.uk/societygov/engineeringethics/pdf/Autonomous_Systems_Report_09.pdf (London: Royal Academy of Engineer, 2009), 2.

21. John Markoff, "Google Cars Drive Themselves, in Traffic," *New York Times,* October 9, 2010, http://www.nytimes.com/2010/10/10/science/10google.html?_r=0.

22. Joanna J. Bryson, "Why Robot Nannies Probably Won't Do Much Psychological Damage," *Interaction Studies* 11:2 (2010): 196–200.

23. As the next chapter discusses more thoroughly, the states that have begun to address autonomous cars have stated that the "operator" of the cars—that is, the person who turns the car on—will be liable for damages caused by the car. However, as I mention above, this does not represent the best public policy regarding these cars. Most people will use them—will be driven by them—without monitoring them the way these early laws assume. We will eventually think assigning liability for damages caused by an autonomous car to the person who turned the car on is as logical as assigning liability for damages caused by an elevator to the person who pushed the floor button.

24. See George S. Cole, "Tort Liability for Artificial Intelligence and Expert Systems," *Computer/Law Journal* 10:2 (April 1990): 213–30.

25. Dobbs and Hayden, 303–7.

26. See Cole, 174–75.

27. Dobbs and Hayden, 916–17.

28. Katherine J. Herrmann, "Cybersurgery: The Cutting Edge," *Rutgers Computer and Technology Law Journal* 32:2 (2006): 305.

29. See "The Kindness of Strangers," *Babbage—Science and Technology* (blog), *The Economist,* January 18, 2012, http://www.economist.com/blogs/babbage/2012/01/surgical-robots; Herrmann, 297–98; Thomas R. McLean, "Cybersurgery: Innovation or a Means to Close Community Hospitals and Displace Physicians?" *John Marshall Journal of Computer and Information Law* 20:4 (Summer 2002): 539.

30. McLean, 506–8.

31. Anthony R. Lanfranco, et al., "Robotic Surgery: A Current Perspective." *Annals of Surgery* 239:1 (January 2004): 15.

32. Justin M. Albani, "The Role of Robotics in Surgery: A Review," *Missouri Medicine* 104:2 (March/April 2007): 166.

33. Brian Davies, "Robotic Surgery: From Autonomous Systems to Intelligent Tools" (transcript of lecture, Institution of Mechanical Engineers, London, July 2007), 1.

34. Lanfranco, et al., 15; John B. Malcolm, Michael D. Fabrizio, and Paul F. Schellhammer, "Witnessing the Transition of Open Robotic Surgery," in A. K. Hemal and M. Menon, eds., *Robotics in Genitourinary Surgery* (London: Springer-Verlag, 2011), 119.

35. "Unimate Puma Series 500 Industrial Robot," Unimation, promotional booklet, http://www.antenen.com/htdocs/downloads/files/files_dl/puma560.pdf (May 1984).

36. Jeremy Pearce, "George C. Devol, Inventor of Robot Arm, Dies at 99," *New York Times,* August 15, 2011. http://www.nytimes.com/2011/08/16/business/george-devol-developer-of-robot-arm-dies-at-99.html?_r=2&partner=rss&emc=rss&.

37. Lisa Nocks, *The Robot: The Life Story of a Technology* (Westport, CT: Greenwood Press, 2007), 69; Roland Menassa, "Robonaut2 and Next Generation Industrial Robots," http://www.robobusiness.com/images/uploads/CS04_Robonaut2_and_Next_Generation_Industrial_Robots.pdf (presentation outline, IEEE 12 International Conference on Intelligent Autonomous Systems, Jeju Island, South Korea, June 26–29, 2012), 5.

38. Nocks, p. 69.

39. Ferdinando Rodriguez and Brian Davies, "Robotic Surgery: From Autonomous Systems to Intelligent Tools," *Robotica* 28: Special Issue 2 (March 2010): 163.

40. Albani, 166–67.

41. Lanfranco, 15.

42. Malcolm, Fabrizio, and Schellhammer, 119.

43. Priya Ganapati, "Surgical Robots Operate with Precision," *Wired,* September 11, 2009, http://www.wired.com/gadgetlab/2009/09/surgical-robots/.

44. Joanne Pransky, "ROBODOC—Surgical Robot Success Story," *Industrial Robot* 24:3 (1997), 231–32; Lanfranco, et al., 15; Peter Kazanzides, e-mail interview with author, April 15 & 27, 2013.

45. Pransky, 231.

46. Rodriguez and Davies, 163.

47. Albani, 167

48. "Sutter General Performs First Hip Replacement with ROBODOC Surgical System after FDA Clearance." *News Medical*, May 29, 2010, http://www.news-medical.net/news/20100529/Sutter-General-performs-first-hip-replacement-with-ROBODOC-Surgical-System-after-FDA-clearance.aspx; "ROBODOC Professionals Page," Robodoc—Curexo Technology Corporation, http://www.robodoc.com/professionals.html.

49. Davies, 2–3.

50. Rodriguez and Davies, 164.

51. M. Jakopec, et al., "The First Clinical Application of a 'Hands-On' Robotic Knee Surgery System," *Computer Aided Surgery* 6:6 (2001): 329–39.

52. Rodriguez and Davies, 164.

53. "Robot Assisted Surgery More Accurate Than Conventional Surgery," Imperial College, press release, February 8, 2006, http://www.imperial.ac.uk/college.asp?P=7449.

54. Rodriguez and Davies, 165.

55. Kristen Gerencher, "Robots as Surgical Enablers." *MarketWatch* (blog), *Wall Street Journal*, February 3, 2005, http://www.marketwatch.com/story/a-fascinating-visit-to-a-high-tech-operating-room?dist=msr_2; Barnaby J. Feder, "Prepping Robots to Perform Surgery," *New York Times,* May 4, 2008, http://www.nytimes.com/2008/05/04/business/04moll.html?pagewanted=2&_r=0; "Regulatory Clearance," Intuitive Surgical, http://www.intuitivesurgical.com/specialties/regulatory-clearance.html.

56. Lanfranco, 16.

57. Ibid., 16–17.

58. Albani, 167.

59. Satyam Kalan, et al., "History of Robotic Surgery," *Journal of Robotic Surgery* 4:3 (September 2010): 144; Albani, 167; Feder, "Prepping Robots to Perform Surgery."

60. Lanfranco, 15.

61. Albani, 167; Feder, "Prepping Robots to Perform Surgery."

62. Satyam, 144–45; Albani, 167.

63. Albani, 167.

64. S. B. Jones and D. B. Jones, "Surgical Aspects and Future Developments in Laparoscopy," *Anesthesiology Clinics of North America* 19:1 (March 2001): 107–24.

65. V. B. Kim, et al., "Early Experience with Telemanipulative Robot-Assisted Laparoscopic Cholecystectomy Using Da Vinci," *Surgical Laparoscopy Endoscopy Percutaneous Techniques* 12:1 (February 2002): 34–40; K. H. Fuchs, "Minimally Invasive Surgery," *Endoscopy* 23:2 (February 2002): 154–59; J. D. Allendorf, et al., "Postoperative Immune Function Varies Inversely with the Degree of Surgical Trauma in a Murine Model." *Surgical Endoscopy* 11:5 (May 1997): 427–30.

66. Lanfranco, 15–16.

67. S. M. Prasad, et al., "Prospective Clinical Trial of Robotically Assisted Endoscopic Coronary Grafting with 1 Year Follow-Up," *Annals of Surgery* 233:6 (June 2001): 725–32.

68. "The Kindness of Strangers," January 18, 2012.

69. "Sensei X Robotic System," Hansen Medical Inc., http://www.hansenmedical.com/us/products/ep/sensei-robotic-catheter-system.php.

70. "The Lab," Stereotaxis, http://www.stereotaxis.com/physicians/the-lab/.

71. J. R. Adler, et al., "The Cyberknife: A Frameless Robotic System for Radiosurgery," *Stereotactic and Functional Neurosurgery* 69:1–4 (Part 2) (1997): 124–28.

72. Koren, "How Raven, the Open-Source Surgical Robot, Could Change Medicine."

73. Peter Kazanzides, e-mail interview with author, April 15, 2013.

74. Ibid.

75. Feder, "Prepping Robots to Perform Surgery."

76. *Mracek v. Bryn Mawr Hosp.*, 610 F.Supp. 2d 401, 402–3 (E.D. Pa. 2009) (hereinafter referred to as *Mracek I*).

77. *Mracek I*, 610 Supp. 2d at 403.

78. *Mracek v. Bryn Mawr Hospital*, 363 F. App'x. 925, 926 (3d Cir. 2010) (hereinafter referred to as *Mracek II*).

79. *Mracek I*, 610 F.Supp. 2d at 407.

80. *Mracek II*, at 925.

81. *Mracek I*, 610 F.Supp. 2d at 403.

82. *Mracek II*, at 926, n. 1.

83. *Mracek I*, 610 F.Supp. 2d at 405; Margo Goldberg, "The Robotic Arm Went Crazy! The Problem of Establishing Liability in a Monopolized Field," *Rutgers Computer and Technology Law Journal* 38:2 (2012): 246.

84. *Mracek I*, 610 F.Supp. 2d at 405–6; see Goldberg, 248.

85. See Goldberg, 248.

86. Ibid., 248–49.

87. *Mracek I*, 610 F.Supp. 2d at 405–6.

88. *Gagliano v. Kaouk*, 2012-Ohio-1047 (Court of Appeals of Ohio, 2012).

89. *Williams v. Desperito*, C.A. N09C-10-164-CLS (Superior Court of Delaware, 2011).

90. See Cole, 208.

Chapter 3

1. Bryan A. Garner, ed., *Black's Law Dictionary*, 9th ed. (St. Paul, MN: West Publishing, 2009), 1398.

2. David S. Landes, *The Wealth and Power of Nations* (New York: W.W. Norton, 1998), 186.

3. Harold D. Woodman, "Economy from 1815 to 1860," in Glenn Porter, ed., *Encyclopedia of American Economic History*, vol. 1 (New York: Charles Scribner's Sons, 1980), 80–81.

4. Ibid.

5. Henry Clay, "On Domestic Manufactures" (delivered in the Senate of the United States, April 6, 1810), in Daniel Mallory, ed., *The Life and Speeches of the Hon. Henry Clay* (New York: Robert P. Bixby & Co., 1843), 196.

6. William Miller, *A New History of the United States* (New York: Dell Publishing, 1962), 163.

7. Albro Martin, "Economy from Reconstruction to 1914," in Glenn Porter, ed., *Encyclopedia of American Economic History*, vol. 1 (New York: Charles Scribner's Sons, 1980), 107.

8. Stephen M. Salsbury, "American Business Institutions before the Railroad," in Glenn Porter, ed., *Encyclopedia of American Economic History*,

vol. 2 (New York: Charles Scribner's Sons, 1980), 615–16; Lewis C. Solmon and Michael Tierney, "Education," in Glenn Porter, ed., *Encyclopedia of American Economic History*, vol. 3 (New York: Charles Scribner's Sons, 1980), 1015–16.

9. Martin, 107.

10. Harry N. Scheiber, "Law and Political Institutions," in Glenn Porter, ed., *Encyclopedia of American Economic History*, vol. 2 (New York: Charles Scribner's Sons, 1980), 502.

11. Solmon and Tierney, 1015–16; Arthur M Johnson, "Economy since 1914," in Glenn Porter, ed., *Encyclopedia of American Economic History*, vol. 1 (New York: Charles Scribner's Sons, 1980), 117.

12. Johnson, 117.

13. Ibid.

14. Thomas K. McCraw, "Regulatory Agencies." in Glenn Porter, ed., *Encyclopedia of American Economic History*, vol. 1 (New York: Charles Scribner's Sons, 1980), 803–4; see Johnson, 127.

15. Johnson, 127–28.

16. See F. Patrick Hubbard, "Regulation of and Liability for Risks of Physical Injury from 'Sophisticated Robots,'" http://robots.law.miami.edu/wp-content/uploads/2012/01/Hubbard_Sophisticated-Robots-Draft-1.pdf (paper presented as a work-in-progress at We Robot Conference, University of Miami School of Law, April 21–22, 2012): 1. Motley notes that regulations must balance fostering innovation in robots (i.e., maximizing the benefits) with providing a fair and efficient allocation of risks (i.e., minimizing the potential harms).

17. See Farhad Manjoo, "Will Robots Steal Your Job?" *Slate*, September 26, 2011, http://www.slate.com/articles/technology/robot_invasion/2011/09/will_robots_steal_your_job_2.single.html.

18. David Strickland, "Autonomous Vehicle Seminar," http://www.nhtsa.gov/staticfiles/administration/pdf/presentations_speeches/2012/Strickland-Autonomous_Veh_10232012.pdf (lecture, Washington, DC. October 23, 2012).

19. "U.S. Government to Begin Process toward Proposing Standards for Autonomous Cars," Associated Press, October 23, 2012, http://www.nydailynews.com/autos/gov-work-proposing-standards-robo-cars-article-1.1190286.

20. "Volvo Calls for Federal Regulations for Autonomous Vehicles," *Traffic Technology Today*, October 25, 2012, http://www.traffictechnologytoday.com/news.php?NewsID=43994.

21. Douglas Newcomb, "How the Feds Will Regulate Autonomous Cars," *MSN*, October 30, 2012, http://editorial.autos.msn.com/blogs/autosblogpost.aspx?post=6242f1c1-c786-4b6e-9035-d6b575cabc45.

22. See Home Equity Loan Consumer Protection Act, 15 U.S.C. §§ 1637 and 1647; Fair Credit Billing Act, 15 U.S.C. §§ 1666-1666j; The Do-Not Call Registry Act of 2003, 15 U.S.C. § 6151.

23. Tom Vanderbilt, "The Real da Vinci Code," *Wired*, November 2004, http://www.wired.com/wired/archive/12.11/davinci.html.

24. Tom Vanderbilt, "Autonomous Cars through the Ages," *Wired,* February 6, 2012, http://www.wired.com/autopia/2012/02/autonomous-vehicle-history/?pid=1580&viewall=true.

25. "The Original Futurama," *Wired,* November 27, 2007, http://www.wired.com/entertainment/hollywood/magazine/15-12/ff_futurama_original.

26. Vanderbilt, "Autonomous Cars through the Ages."

27. Jameson M. Wetmore, *Driving the Dream,* Consortium for Science, Policy & Outcomes, Arizona State, https://docs.google.com/viewer?a=v&q=cache:acKQ4CCb7MIJ:www.cspo.org/documents/article_Wetmore-DrivingTheDream.pdf+&hl=en&gl=us&pid=bl&srcid=ADGEESh-3ZruXZdm3cxohIOmDZGCkSqgSqeXR70NB6Z0F-YGs1EbzCZt8xZVKJaF8J4Iwi4xVd-8QTY6HTWnAgLvwpPN36RPtIstMz4OwyW52V5opxLEUnNn-gvU_WSkm4ID67_Sa8ez&sig=AHIEtbTqF_hYcYQZ8sUb0nM22Hqb-juirQ, 2.

28. Wetmore, 4–5.

29. Ibid., 5.

30. Ibid., 7.

31. "An Automatically Guided automobile Cruised along a One-Mile Check Road at General Motors Technical Center Today," General Motors Corporation, press release, February 14, 1958.

32. Wetmore, 10.

33. Ibid., 2.

34. Douglas W. Gage, "UGV History 101: A Brief History of Unmanned Ground Vehicle (UGV)," http://www.dtic.mil/cgi-bin/GetTRDoc?Location=U2&doc=GetTRDoc.pdf&AD=ADA422845, *Unmanned Systems* 13:3 (Summer 1995): 1.

35. Gage, 2–3.

36. Vanderbilt, "Autonomous Cars through the Ages."

37. Les Earnest, "Stanford Cart," Stanford University, December 2012, http://www.stanford.edu/~learnest/cart.htm.

38. Vanderbilt, "Autonomous Cars through the Ages."

39. Ernst D. Dickmanns, "Vehicles Capable of Dynamic Vision," *IJCAI '97—Proceedings of the Fifteenth International Joint Conference on Artificial Intelligence,* vol. 2 (San Francisco: Morgan Kaufmann Publishers, 1997), 1577; Tom Vanderbilt, "Autonomous Cars through the Ages."

40. Fourth Conference on Artificial General Intelligence, introductory remarks to "Dynamic Vision as Key Element for AGI," August 4, 2011, http://www.youtube.com/watch?v=YZ6nPhUG2i0; Tom Vanderbilt, "Autonomous Cars through the Ages."

41. Vanderbilt, "Autonomous Cars through the Ages."

42. Gage, 3.

43. Dickmanns, 1589.

44. Vanderbilt, "Autonomous Cars through the Ages."

45. Dickmanns, 1589.

46. No Hands Across America General Information, http://www.cs.cmu.edu/afs/cs/usr/tjochem/www/nhaa/general_info.html.

47. No Hands Across America, http://www.cs.cmu.edu/afs/cs/usr/tjochem/www/nhaa/nhaa_home_page.html.

48. No Hands Across America Journal, July 28–30, 1995, http://www.cs.cmu.edu/afs/cs/usr/tjochem/www/nhaa/Journal.html.

49. Bruce G. Buchanan, "A (Very) Brief History of Artificial Intelligence," *AI* 26:4 (Winter 2005): 60, n. 2.

50. "DARPA Grand Challenge—Leveraging American Ingenuity," presentation outline, http://archive.darpa.mil/grandchallenge04/overview_pres.pdf.

51. Guna Seetharaman, Arun Lakhotia, and Erik Philip Blasch, "Unmanned Vehicles Come of Age: The DARPA Grand Challenge," *Computer*, December 2006, 28–29.

52. John Markoff, "Google Cars Drive Themselves in Traffic." *New York Times,* October 9, 2010, http://www.nytimes.com/2010/10/10/science/10google.html?pagewanted=1&partner=rss&emc=rss&_r=0.

53. Abby Haglage, "Google, Audi, Toyota, and the Brave New World of Driverless Cars," *Daily Beast,* January 16, 2013, http://www.thedailybeast.com/articles/2013/01/16/google-audi-toyota-and-the-brave-new-world-of-driverless-cars.html.

54. John Naughton, "Google's Self-Guided Car Could Drive the Next Wave of Unemployment," *The Guardian,* September 29, 2012, http://www.guardian.co.uk/technology/2012/sep/30/google-self-driving-car-unemployment. In August 2011, a Google Car caused a five-car accident, but the company reported that the car was in manual mode at the time of the accident. Matt Weinberger, "Google Driverless Car Causes Five-Car Crash," *ZD Net,* August 8, 2011, http://www.zdnet.com/blog/google/google-driverless-car-causes-five-car-crash/3211.

55. Sebastian Thrun, "What We're Driving At," *Google Official Blog,* October 9, 2010, http://googleblog.blogspot.com/2010/10/what-were-driving-at.html.

56. Alex Taylor III, "Is Google Motors the New GM?" *Money,* May 24, 2011, http://money.cnn.com/2011/05/23/autos/google_driverless_cars.fortune/index.htm.

57. Tom Vanderbilt, "Let the Robot Drive," *Wired,* January 12, 2012, http://www.wired.com/magazine/2012/01/ff_autonomouscars/all/1.

58. A wonderful resource for tracking the development of legislation addressing autonomous cars is the "Automated Driving: Legislative and Regulatory Action" article from CyberWiki, http://cyberlaw.stanford.edu/wiki/index.php/Automated_Driving:_Legislative_and_Regulatory_Action

59. Florida House Bill 1207 (Ch. 2012-111), § 1(2); California Acts, Chapter 570 of 2012, 2011-2012, § 1(c).

60. Florida House Bill 1207 (Ch. 2012-111), § 5(2).

61. District of Columbia, Automated Vehicle Act of 2012, § 4.
62. However, according to the CyberWiki article, "Automated Driving: Legislative and Regulatory Action," the California legislature struck language regarding the liability of original manufacturers from the legislation.
63. Nevada Assembly Bill 511, 2012, § 2.
64. Arizona House Bill 2679, 2012, § 1.
65. Hawaii House Bill 2238, 2012, § 1.
66. New Jersey Assembly, No. 2757, 2012, § 2.
67. Oklahoma House Bill 3007, 2012, § 1.
68. District of Columbia, Automated Vehicle Act of 2012, § 3.
69. Florida House Bill 1207 (Ch. 2012-111), § 3.
70. California Acts, Chapter 570 of 2012, 2011-2012, § 2.
71. NAC Chapter 482A.020.
72. NAC Chapter 482A.030.
73. California Acts, Chapter 570 of 2012, 2011-2012, § 2.
74. Nevada Assembly Bill 511, 2012, § 5; NAC Chapter 482A.100-180.
75. Florida House Bill 1207 (Ch. 2012-111), § 5.
76. California Acts, Chapter 570 of 2012, 2011-2012, § 2.
77. Florida House Bill 1207 (Ch. 2012-111), § 4.
78. NAC Chapter 482A.190.
79. NAC Chapter 482A.190; California Acts, Chapter 570 of 2012, 2011-2012, § 2; Florida Chapter 2012-111, House Bill 1207, § 4.
80. California Acts, Chapter 570 of 2012, 2011-2012, § 2.
81. Ibid.
82. Arizona House Bill 2679, 2012; California Acts, Chapter 570 of 2012, 2011-2012; District of Columbia, "Automated Vehicle Act of 2012"; Florida House Bill 1207 (Ch. 2012-111); Hawaii House Bill 2238, 2012; New Jersey Assembly, No. 2757, 2012; Nevada Assembly Bill 511, 2012; Oklahoma House Bill 3007, 2012.
83. *See* Yvette Joy Liebesman, "The Wisdom of Legislating for Anticipated Technological Advances," *John Marshall Review of Intellectual Property* 10:1 (2010): 154–81.
84. In the United States, the Uniform Commercial Code has been passed in all 50 states and seeks to provide consistent rules and laws for commercial transactions regardless of the state. See Cornell University Law School's UCC resource page for the text and more information: http://www.law.cornell.edu/ucc/ucc.table.html.
85. The Uniform Trust Code is the model law in the United States addressing the creation and administration of trusts, particularly in their role as an estate-planning tool. As of 2012, 24 states had adopted the entirety of the UTC or some portion of it. Please see the Uniform Law Commission's website for more information: http://www.uniformlaws.org/Act.aspx?title=Trust%20Code.
86. The Uniform Controlled Substances Act is the model law in the United States addressing controlled substances. As of 2012, 30 states had adopted all

or part of it. Please see the Uniform Law Commission's website for more information: http://uniformlaws.org/Act.aspx?title=Controlled%20Substances%20Act.

87. Florida Chapter 2012-111, House Bill 1207, § 4.

88. NAC Chapter 482A.050.

89. California Acts, Chapter 570 of 2012, 2011-2012, § 2.

90. Arizona House Bill 2679, 2012, § 1. Hawaii, Oklahoma, and New Jersey have similar language.

91. "Info on 3.9M Citigroup customers lost," *CNNMoney,* June 6, 2005, http://money.cnn.com/2005/06/06/news/fortune500/security_citigroup/.

92. Jonathan Krim and Michael Barbaro, "40 Million Credit Card Numbers Hacked," *Washington Post,* June 18,, 2005, http://www.washingtonpost.com/wp-dyn/content/article/2005/06/17/AR2005061701031.html.

93. David Stout and Tom Zeller Jr., "Vast Data Cache about Veterans Is Stolen," *New York Times,* May 22, 2006, http://www.nytimes.com/2006/05/23/washington/23identity.html.

94. Jaikumar Vijayan, "TJX Data Breach: At 45.6M Card Numbers, It's the Biggest Ever," *Computer World,* March 29, 2007, http://www.computerworld.com/s/article/9014782/TJX_data_breach_At_45.6M_card_numbers_it_s_the_biggest_ever?taxonomyId=17&pageNumber=1.

95. "Chronology of Data Breaches, Security Breaches 2005–Present," Privacy Rights Clearinghouse, http://www.privacyrights.org/data-breach, last updated on January 8, 2013.

96. E. Scott Reckard and Joseph Menn, "Insider Stole Countrywide Applicants' Data, FBI Alleges," August 2, 2008, *Los Angeles Times,* http://articles.latimes.com/2008/aug/02/business/fi-arrest2.

97. "Chronology of Data Breaches, Security Breaches 2005–Present."

98. Byron Acohido, "Hackers Breach Heartland Payment Credit Card System," *USA Today,* January 23, 2009, http://usatoday30.usatoday.com/money/perfi/credit/2009-01-20-heartland-credit-card-security-breach_N.htm.

99. *United States v. Gonzalez,* indictment, 3 (United States District Court District of New Jersey), accessed at http://www.wired.com/images_blogs/threatlevel/2009/08/gonzalez.pdf.

100. "Video Game Company Valve Notifies Its Gamers of Data Breach," *Alertsec Xpress* (blog), November 15, 2011, http://blog.alertsec.com/2011/11/video-game-company-valve-notifies-its-gamers-of-data-breach/.

101. "Chronology of Data Breaches, Security Breaches 2005–Present."

102. For a comprehensive list, please see the "Chronology of Data Breaches, Security Breaches 2005–Present" at http://www.privacyrights.org/data-breach.

103. Charles Duhigg, "How Companies Learn Your Secrets," *New York Times,* February 12, 2012, http://www.nytimes.com/2012/02/19/magazine/shopping-habits.html?pagewanted=7&_r=1&hp.

104. Ibid.

105. Hubbard, 34.

106. See New Hampshire Midwifery Council, the administrative rules for which can be viewed here: http://www.gencourt.state.nh.us/rules/state_agencies/mid.html.

107. See California Board of Guide Dogs for the Blind, http://www.guidedogboard.ca.gov/.

108. See Texas Council on Purchasing from Persons with Disabilities, http://www.tcppd.state.tx.us/.

Chapter 4

1. *Schall v. Martin*, 467 U.S. 253, 268 (1984).

2. Bryan A. Garner, ed., *Black's Law Dictionary*, 9th ed. (St. Paul, MN: West Publishing, 2009), 235, 322.

3. Megan F. Chaney, "Keeping the Promise of *Gault*: Requiring Post-Adjudicatory Juvenile Defenders," *Georgetown Journal on Poverty Law and Policy* 19:3 (Summer 2012): 367.

4. See *Schall*, 467 U.S. at 268.

5. *Meyer v. Nebraska*, 262 U.S. 390, 399–400 (1923).

6. *Santosky v. Kramer*, 455 U.S. 745, 766 (1982).

7. See *West Virginia State Board of Education v. Barnette*, 319 U.S. 624, 629, 632 (1943).

8. *Tinker v. Des Moines Independent Community School District*, 393 U.S. 503 (1969).

9. *Schall* , 467 U.S. at 263.

10. See Barry C. Feld, *Cases and Materials on Juvenile Justice Administration* (St. Paul, MN: West Publishing, 2004), 202.

11. Feld, 3; Chaney, 367.

12. See Mark E. Sullivan, "Military Custody Twists & Turns," *Family Advocate* 28:2 (Fall 2005): 23–24.

13. Linda A. Chapin, "Out of Control? The Uses and Abuses of Parental Liability Laws to Control Juvenile Delinquency in the United States," *Santa Clara Law Review* 37:3 (1997): 632.

14. Ibid., 630.

15. Ibid., 632–33.

16. Joseph E. Brick, "Non-Custodial Parent's Liability for Tortious Timmy's Delinquent Acts," *Journal of Juvenile Law* 27 (2006): 84.

17. In an effort to avoid unnecessary confusion, I will not discuss conservatorships, which are frequently mistaken for guardianships and vice versa. Typically, in a conservatorship, one person assumes the right to make decisions about another person's finances, whereas in a guardianship one person assumes the right to make decisions about another person. This chapter addresses AI that can physically care for and look after people, so conservatorships are not relevant. Having said that, I have no doubt that Merrill Lynch is attempting to reprogram HAL to manage your trust fund.

18. Andrew H. Hook and Lisa V. Johnson, "The Uniform Power of Attorney Act," *Real Property, Trust and Estate Law Journal* 45:2 (Summer 2010): 285–86.

19. Erica F. Wood, *State Level Adult Guardianship Data: An Exploratory Survey*, http://www.ncea.aoa.gov/ncearoot/main_site/pdf/publication/guardian shipdata.pdf (Washington, DC: American Bar Association Commission on Law and Aging, 2006), 8–9.

20. Uniform Law Commission, "Uniform Power of Attorney Act," http://www.uniformlaws.org/Act.aspx?title=Power%20of%20Attorney, § 102(5) (accessed April 29, 2013).

21. See Hook and Johnson, 285.

22. Wood, 5.

23. Barbara A. Cohen, Barbara Oosterhout, and Susan P. Leviton, "Tailoring Guardianship to the Needs of Mentally Handicapped Citizens," *Maryland Law Forum* 6:3 (1976): 92.

24. Lord Coke, *Beverley's Case*, 76 Eng. Rep. 1118, 1122 (K.B. 1603).

25. Cohen, Oosterhout, and Leviton, 92.

26. Michael D. Casasanto, Mitchell Simon, and Judith Roman, "A Model Code of Ethics for Guardians," *Whittier Law Review* 11:3 (1989): 566; see Hook and Johnson, 297–98.

27. See Hook and Johnson, 300.

28. Martha E. Pollack, et al., 2002 "Pearl: A Mobile Robotic Assistant for the Elderly," *AAAAI Technical Report*, WS-02-02, http://www.aaai.org/Papers/ Workshops/2002/WS-02-02/WS02-02-013.pdf, 2 (hereinafter referred to as Pollack 1).

29. Robert H. Friedman, "Automated Telephone Conversations to Assess Health Behavior and Deliver Behavioral Interventions," *Journal of Medical Systems* 22:2 (1998): 95–102.

30. Bodil Jönsson and Arne Svensk, "Isaac: A Personal Digital Assistant for the Differently Abled," in I. Placencia Porrero and R. Puig de la Bellacasa, eds., *The European Context for Assistive Technology* (Amsterdam: IOS Press, 1995), 356–61.

31. Neil Hersh and Larry Treadgold, "NeuroPage: The Rehabilitation of Memory Dysfunction by Prosthetic Memory and Cueing," *NeuroRehabilitation* 4 (1994): 187–97.

32. Pollack, 2; Friedman, 96.

33. Martha E. Pollack, "Planning Technology for Intelligent Cognitive Orthotics," *American Association for Artificial Intelligence* (2002), http://www .cs.cmu.edu/~flo/papers/umich/AIPS-02Pollack.pdf, 2 (hereinafter referred to as Pollack 2).

34. See James P. Turley, "The Use of Artificial Intelligence in Nursing Information Systems," *Informatics in Healthcare Australia*, May 1993, http:// www.project.net.au/hisavic/hisa/mag/may93/the.htm.

35. See Turley.

36. Pollack 1, 1.

37. Diane Stresing, "Artificial Caregivers Improve on the Real Thing," *TechNewsWorld*, August 30, 2003, http://www.technewsworld.com/story/31465.html.

38. Pollack 1, 2–3.

39. Pollack 2, 8.

40. Stresing, "Artificial Caregivers Improve on the Real Thing."

41. Joelle Pineau, et al., "Towards Robotic Assistants in Nursing Homes: Challenges and Results," *Robotics and Autonomous Systems* 42 (2003): 272.

42. Pollack 2, 8.

43. Janet Boivin and Scott Williams, "Robots Become Nurses' Valuable Assistants," *Nurse.com*, March 10, 2008, http://news.nurse.com/apps/pbcs.dll/article?AID=/20080310/NATIONAL01/80307014/-1/frontpage.

44. Boivin and Williams, "Robots Become Nurses' Valuable Assistants."

45. Stresing, "Artificial Caregivers Improve on the Real Thing."

46. Richard Lai, "Family Nanny Robot Is Just Five Years and $1,500 Away from Being Your New Best Friend," *Engadget,* April 30, 2010, http://www.engadget.com/2010/04/30/family-nanny-robot-is-just-five-years-and-1-500-away-from-being/.

47. "Robotic Nursing Assistant," Small Business Innovation Research. Small Business and Award Information—RE2, Inc., http://www.sbir.gov/sbirsearch/detail/290061.

48. Boivin and Williams, "Robots Become Nurses' Valuable Assistants." Pedersen changed positions when she became a Certified Holistic Health Coach in January 2013 and currently serves as the part-time Marketing Director of RE2, Inc.

49. Chih-Hung King, Tiffany L. Chen, Advait Jain, and Charles C. Kemp, "Towards an Assistive Robot That Autonomously Performs Bed Baths for Patient Hygiene," *Georgia Institute of Technology,* http://www.hsi.gatech.edu/hrl/pdf/iros10_auto_clean.pdf.

50. Noel Sharkey, "The Ethical Frontiers of Robotics," *Science* 322:5909 (December 19, 2008), 1800–1801. Sharkey also warns of the ethical problems in permitting robots to care for the elderly and perform military strikes.

51. Sharkey, "The Ethical Frontiers of Robotics," 1800.

52. Joanna E. Bryson, "Why Robot Nannies Probably Won't Do Much Psychological Damage," *Interaction Studies* 11:2 (2010), 196–200.

53. Brandon Keim, "I, Nanny: Robot Babysitters Pose Dilemma," *Wired*, December 18, 2008. http://www.wired.com/wiredscience/2008/12/babysittingrobo/.

54. "PaPeRo Product Page," NEC, http://www.nec.co.jp/products/robot/en/index.html.

55. "PaPeRo Functions Page," NEC, http://www.nec.co.jp/products/robot/en/functions/index.html.

56. "Updated: Robot Babysitters Hit Japanese Stores," *TechRadar,* March 26, 2008, http://www.techradar.com/us/news/world-of-tech/future-tech/updated-robot-babysitters-hit-japanese-stores-272406.

57. Susan Karlin, "A Scientist Creates Robots That Help Children," *IEEE Spectrum*, February 2010, http://spectrum.ieee.org/robotics/humanoids/caregiver-robots.

58. Sarah Shemkus, "Students Seek Connection with Robots," *Boston Globe*, April 19, 2013, http://www.bostonglobe.com/business/2013/04/18/haverhill-school-tests-french-made-robots-way-teach-children-with-learning-disabilities/yB2GOwEKcRnm0eXFxSO7iO/story.html.

59. Elizabeth Kazakoff, e-mail interview with author, April 23, 2013.

60. Susannah Palk, "Robot Teachers Invade South Korean Classrooms," *CNN.com*, October 22, 2010, http://www.cnn.com/2010/TECH/innovation/10/22/south.korea.robot.teachers/index.html.

61. Palk, "Robot Teachers Invade South Korean Classrooms."

62. Chloe Albanesius, "Honda Unveils Faster, Smarter ASIMO 'Humanoid' Robot," *PC*, November 8, 2011, http://www.pcmag.com/article2/0,2817,2396071,00.asp.

63. "Robot Babysitter," *Heather and Randy's Family Blog*, March 24, 2011, http://heatherandrandyfam.blogspot.com/2011/03/robot-babysitter.html.

64. "Robot Babysitter," *Heather and Randy's Family Blog*.

65. "Idea #3—Robot Nanny," *HCI 2 Blog*, February 12, 2007, http://hci2blog2007.blogspot.com/2007/02/idea-3-robot-nanny.html.

66. Timothy Bush, *Benjamin McFadden and the Robot Babysitter* (New York: Crown, 1998).

67. King, Chen, Jain, and Kemp.

68. Pollack 1.

69. Kerstin Roger, et al., "Social Commitment Robots and Dementia," *Canadian Journal on Aging* 31:1 (March 2012).

70. See chapter 2 for a more complete discussion of these concepts.

Chapter 5

1. John Markoff, "Skilled Work Without the Worker," *New York Times*, August 18, 2012, http://www.nytimes.com/2012/08/19/business/new-wave-of-adept-robots-is-changing-global-industry.html?pagewanted=all&_r=0.

2. *Village of Euclid v. Ambler Realty Co.*, 272 U.S. 365 (1926).

3. *Hunter v. City of Pittsburgh*, 207 U.S. 161, 178 (1907); Harvey Walker, *Federal Limitations upon Municipal Ordinance Making Power* (Columbus: Ohio State University Press, 1929), 2–3.

4. Gerald E. Frug, "The City as a Legal Concept," *Harvard Law Review* 93:6 (April 1980): 1059.

5. Richard Briffault, "Our Localism: Part I—The Structure of Local Government Law," *Columbia Law Review* 90:1 (January 1990): 111–12.

6. Briffault, 112.

7. Walker, 4.

8. Walker, 4; Frug, 1083–84.

9. Frug, 1092, n. 135. Jennifer Levin's *The Charter Controversy in the City of London 1660–1688, and Its Consequences* has a thorough discussion of this legal dispute.

10. Frug, 1090–95.

11. Ibid., 1096–97.

12. Lynn A. Baker and Daniel B. Rodriguez, "Constitutional Home Rule and Judicial Scrutiny," *Denver University Law Review* 86:5 (2009): 1340, n. 19.

13. See Frug, 1105–9; Briffault, 6–8.

14. Baker and Rodriguez, 1340.

15. Walker, 12.

16. Briffault, 19.

17. See Walker, 12.

18. Baker and Rodriguez, 1409.

19. Ibid., 1410.

20. Briffault, 18–39.

21. Ibid., 17–18.

22. Jesse Dukeminier and James E. Krier, *Property*, 5th ed. (New York: Aspen Publishers, 2002), 952.

23. Lewis Mumford, *The City in History: Its Origins, Its Transformations, and Its Prospects* (New York: Harcourt, Brace & World, 1961), 433.

24. Dukeminier and Krier, 952–53.

25. Ibid., 958.

26. Ibid., 959.

27. Brief of Alfred Bettman, *Amici Curiae*, on behalf of the National Conference on City Planning, the National Housing Association, and the Massachusetts Federation of Town Planning Boards, *Village of Euclid v. Ambler Realty* Co. 272 U.S. 365 (1926), 5.

28. Dukeminier and Krier, 959.

29. See Joseph Gordon Hylton, "Prelude to Euclid: The United States Supreme Court and the Constitutionality of Land Use Regulation, 1900–1920." *Washington University Journal of Law and Policy* 3 (2000): 1–37; Dukeminier and Krier, 959.

30. See *Village of Euclid*, 272 U.S. at 386, 395.

31. *Village of Euclid*, 272 U.S. at 384.

32. Ibid. at 391.

33. Robert C. Ellickson, "Alternatives to Zoning: Covenants, Nuisance Rules, and Fines as Land Use Controls," *University of Chicago Law Review* 40:4 (Summer 1973): 692.

34. Although it is not relevant to my brief discussion of zoning ordinances here, it is worth noting that critics of zoning point out that it promotes an unfair distribution of wealth as well as economic and racial segregation. There is a great deal of evidence to support this claim. Low-income and minority groups have attacked exclusionary zoning practices—minimum lot sizes, barring mobile or manufactured homes, etc.—in court as contrary to the obligation every

municipality has to promote public health, safety, and general welfare. Such attacks have met with some success. See *Southern Burlington County NAACP v. Township of Mount Laurel*, 67 N.J. 151 (1975) and *Southern Burlington County NAACP v. Township of Mount Laurel*, 92 N.J. 158 (1983) for further discussion.

35. *Village of Euclid*, 272 U.S. at 380.

36. Johanna Wallén, "The History of the Industrial Robot," *Technical Report from Automatic Control at Linköpings Universitet*, Report No. LiTH-ISY-R-2853, May 8, 2008, http://www.control.isy.liu.se/research/reports/2008/2853.pdf, 5.

37. Kristina Dahlin, "Diffusion and Industrial Dynamics in the Robot Industry," in Bo Carlsson, ed., *Technological Systems and Economic Performance: The Case of Factory Automation* (Norwell, MA: Kluwer Academic Press, 1995), 326.

38. Karl Mathia, *Robotics for Electronics Manufacturing* (Cambridge: Cambridge University Press, 2010), 1.

39. Wallén, "The History of the Industrial Robot," 3.

40. Mathia, 1.

41. International Federation of Robotics, *History of Industrial Robots* (brochure, 2012), http://www.ifr.org/uploads/media/History_of_Industrial_Robots_online_brochure_by_IFR_2012.pdf.

42. Wesley L. Stone, "The History of Robotics," in Thomas R. Kurfees, ed., *Robotics and Automation Handbook* (Boca Raton, FL: CRC Press, 2005), 1–5.

43. Stone, "The History of Robotics," 1-4–1-5.

44. George C. Devol, 1961, Program Article Transfer, U.S. Patent 2,988,237, filed December 10, 1954 and issued June 13, 1961.

45. Stone, "The History of Robotics," 1–5.

46. Ibid.," 1–7.

47. Wallén, "The History of the Industrial Robot," 10.

48. Mathia, 3.

49. Lisa Nocks, *The Robot: The Life Story of a Technology* (Westport, CT: Greenwood Press, 2007), 69.

50. Mathia, 3.

51. International Federation of Robotics, *History of Industrial Robots*; Mathia, 3.

52. International Federation of Robotics, *History of Industrial Robots*.

53. Wallén, "The History of the Industrial Robot," 11.

54. Stone, "The History of Robotics," 1–7.

55. Dahlin, 330.

56. Stone, "The History of Robotics," 10.

57. Ibid., 1–7.

58. Mathia, 3.

59. Chapter 2 discusses this in more detail.

60. Nocks, 69.

61. Mathia, 3–5.

62. Dahlin, 323.

63. Mathia, 4–5.

64. Wallén, "The History of the Industrial Robot," 10–11.

65. Susan W. Sanderson and Brian J. L. Berry, "Robotics and Regional Development," in John Rees, ed., *Technology, Regions, and Policy* (Totowa, NJ: Rowman & Littlefield, 1986), 173.

66. Mathia, 6.

67. International Federation of Robotics, *History of Industrial Robots.*

68. Dahlin, 329; Mathia, 5; International Federation of Robotics, *History of Industrial Robots.*

69. Dahlin, 325–26.

70. Mathia, 7.

71. "Fields of Automation," *The Economist,* December 10, 2009, http://www.economist.com/node/15048711.

72. Yoshisada Nagasaka, et al., "High-Precision Autonomous Operation Using an Unmanned Rice Transplanter," in K. Toriyama, K. L. Heong, and B. Hardy, eds., *Rice Is Life: Scientific Perspectives for the 21 Century—Proceedings of the World Rice Research Conference, Tsukuba, Japan* [CD-ROM], 235–37 .

73. "Fields of Automation," *The Economist,* December 10, 2009.

74. Ibid.

75. "Robotic Platform," The Distributed Robotics Garden, Massachusetts Institute of Technology, people.csail.mit.edu/nikolaus/drg/index.php/robots.

76. Julie Gordon, "Miners Take 'Rail-Veyors' and Robots to Automated Future," *Reuters,* October 28, 2012, http://www.reuters.com/article/2012/10/28/us-mining-technology-idUSBRE89R06C20121028.

77. Emma Bastian, "There's No Canary in These Mines," *Features, Robotics, Technology* (blog), *Science Illustrated,* April 3, 2012, http://www.scienceillustrated.com.au/blog/features/theres-no-canary-in-these-mines.

78. Alexis Madrigal, "Autonomous Robots Invade Retail Warehouses," *Wired,* January 27, 2009, http://www.wired.com/wiredscience/2009/01/retailrobots/.

79. Evelyn M. Rusli, "Amazon.com to Acquire Manufacturer of Robotics," *DealBook* (blog), *New York Times,* March 19, 2012, dealbook.nytimes.com. 2012/03/19/amazon-com-buys-kiva-systems-for-775-million/.

80. "Industries Page," Kiva Systems, http://www.kivasystems.com/industries/.

81. Mao Jing, "Foxconn Halts Recruitment as They Look to Automated Robots," *China Daily USA,* February 20, 2013, http://usa.chinadaily.com.cn/business/2013-02/20/content_16240755.htm.

82. Gregory T. Huang, "Rod Brooks and Rethink Reveal an Industrial Robot for the Masses," *Xconomy,* September 18, 2012, http://www.xconomy.com/boston/2012/09/18/rod-brooks-and-rethink-reveal-an-industrial-robot-for-the-masses/.

83. Christopher Mims, "How Robots Are Eating the Last of America's—and the World's—Traditional Manufacturing Jobs," *Quartz*, February 15, 2013, http://qz.com/53710/robots-are-eating-manufacturing-jobs/.

84. Huang, "Rod Brooks and Rethink Reveal an Industrial Robot for the Masses."

85. Dan Human, "City Reports Rise in Parking Meter Profit, Revenue," *Indianapolis Business Journal*, February 21, 2013, http://www.ibj.com/city-reports-rise-in-parking-meter-profit-revenue/PARAMS/article/39741.

86. Jon Halpern, "Boston Parking Fine Revenue Still Significantly Down from 2010 Peak," *Boston Business Journal*, September 27, 2012, http://www.bizjournals.com/boston/blog/bbj_research_alert/2012/09/boston-parking-violations.html?page=all.

87. Ashley Halsey III, "D.C. Sets Record with Parking Ticket Revenue," *Washington Post*, March 5, 2012, http://articles.washingtonpost.com/2012-03-05/local/35448001_1_ticket-fines-meter-revenue-unpaid-tickets.

88. "Transportation Profile," Los Angeles Department of Transportation, http://ladot.lacity.org/about_transportation_profile.htm.

89. See Joan Nassauer, "The Aesthetic Benefits of Agricultural Land," *Renewable Resources Journal* 7:4 (Winter 1989), 17–18.

90. These are common provisions in zoning ordinance sections governing home offices and home occupations.

91. This is not uncommon language used to govern adult entertainment establishments in zoning ordinances.

92. Daniel Gross, "What's the Toll? It Depends on the Time of Day," *New York Times*, February 11, 2007, http://www.nytimes.com/2007/02/11/business/yourmoney/11view.html.

93. See Randolph Langenbach, "An Epic in Urban Design," *Harvard Alumni Bulletin* 70:12 (April 13, 1968): 19; John R. Mullin, Jeanne H. Armstrong, and Jean S. Kavanagh, "From Mill Town to Mill Town: The Transition of a New England Town from a Textile to a High-Technology Economy," *Journal of the American Planning Association* 52:1 (1986): 47–59.

94. See John Mullin and Zenia Kotval, "Assessing the Future of the New England Mill Town: What Are the Key Factors That Lead to Successful Revitalization?" *Landscape Architecture & Regional Planning Faculty Publication Series*, Paper 22, 1986, http://scholarworks.umass.edu/larp_faculty_pubs/22. Retrieved on 3-1-2013.

95. George Waldo Browne, *The Amoskeag Manufacturing Co. of Manchester, New Hampshire: A History* (Manchester, NH: Amoskeag Manufacturing Company, 1915), 61–62.

96. Browne, 152.

97. "Manchester and the Amoskeag." *New Hampshire Public Television*. http://www.nhptv.org/kn/itv/ournh/ournhtg10.htm (accessed on January 14, 2013.

98. John R. McLane, Jr., *"Judge" McLane: His Life and Times and the McLane Law Firm* (Portsmouth, NH: Peter E. Randall, 1996), 149–66.

99. "Varied Land Use Creates Unique Challenges," *Union Leader*, February 15, 1993; Nancy Meersman, "What Will Become of the Millyard?" *Union Leader*, March 30, 1990.

100. Mullin and Kotval.

101. Alan C. Weinstein and Richard McCleary, "The Association of Adult Businesses with Secondary Effects: Legal Doctrine, Social Theory, and Empirical Evidence," *Cardozo Arts & Entertainment Law Journal* 29:3 (2011): 565–96.

102. *Renton v. Playtime Theatres Inc.*, 475 U.S. 41 (1986).

103. See "Repeal of City Ordinance Leads to Rise in Prostitution; Neighbors Protesting," *KMOV.com*, August 14, 2012, http://www.kmov.com/news/ editors-pick/Repeal-of-city-ordinance-leads-to-rise-in-prostitution-neighbors -protesting-166206286.html.

Chapter 6

1. 533 U.S. 27 (2001).

2. 463 Mass. 790 (2012).

3. 132 S. Ct. 945 (2012).

4. For more information on privacy and Fourth Amendment concerns involving autonomous cars, see Dorothy J. Glancy, "Privacy in Autonomous Vehicles," *Santa Clara Law Review* 52:4 (2012): 1171–1239.

5. Osmond K. Fraenkel, "Concerning Search and Seizure," *Harvard Law Review* 34:4 (1921): 362–63.

6. Lord Camden, *Entick v. Carrington*, 19 How. St. Tr. 1029, 1073 (1765).

7. *Boyd v. United States*, 116 U.S. 616, 624–25 (1886).

8. Thomas Y. Davies, "Recovering the Original Fourth Amendment," *Michigan Law Review* 98:3 (December 1999): 566–67.

9. John Adams to William Tudor, March 29, 1817, in Charles Francis Adams, ed., *The Works of John Adams, Second President of the United States*, vol. 10 (Boston: Little, Brown, and Company, 1856), 248.

10. Fraenkel, 361–62.

11. This statement represents a reconciliation of irreconcilable positions taken by Akhil Reed Amar and Thomas Y. Davies. Amar believes that the Framers wanted to ensure "that all government searches and seizures be reasonable." Akhil Reed Amar, "Fourth Amendment and First Principles," *Harvard Law Review* 107:4 (February 1994): 759. Davies believes that the Framers "aimed the Fourth Amendment precisely at banning Congress from authorizing use of general warrants; they did not mean to create any broad reasonable standard for assessing warrantless searches and arrests." Davies, 724. Both present reasonable and compelling arguments. Because the difference is irrelevant to this book, I have chosen to adopt aspects of each.

12. David E. Steinberg, "The Original Understanding of Unreasonable Searches and Seizures," *Florida Law Review* 56:5 (December 2004): 1071–72.

13. Fraenkel, 361.

14. *Adams v. New York,* 192 U.S. 594 (1904).

15. *Mapp v. Ohio,* 367 U.S. 643, 650, quoting *Wolf v. Colorado,* 338 U.S. 25, 27–29 (1948).

16. *Mapp,* 367 U.S. at 655–57.

17. *Katz v. United States,* 389 U.S. 347 (1967).

18. *Katz,* 389 U.S. at 361 (Harlan, J., concurring).

19. *Ibid.*

20. *Skinner v. Railway Labor Executives Association*, 489 U.S 602, 619 (1989).

21. Ibid. at 616–18.

22. Ibid. at 617.

23. Ibid. at 618–34.

24. *Smith v. Maryland,* 442 U.S. 735, 737 & 742–43 (1979).

25. Steinberg, 1056.

26. *Smith*, 442 U.S. at 743–46.

27. *Kyllo,* 533 U.S. at 29–30.

28. Ibid. at 31.

29. Ibid. at 33.

30. *California v. Ciraolo,* 476 U.S. 207 (1986).

31. *Florida v. Riley*, 488 U.S. 445 (1989).

32. *Dow Chemical Co. v. United States*, 476 U.S. 227 (1986).

33. *Kyllo,* 533 U.S. at 33.

34. Ibid. at 40.

35. *Phifer*, 463 Mass. at 794–95.

36. Ibid. at 790–91. The Supreme Judicial Court also noted that. in addition to alleging that the use of that evidence violated the Fourth Amendment, Phifer also alleged that it violated the Fourteenth Amendment, which made the Fourth Amendment enforceable against the states.

37. *Phifer*, 463 Mass. at 793–97.

38. Ibid. at 797.

39. *Jones*, 132 S. Ct. at 948.

40. Ibid. at 948–49.

41. Ibid. at 952.

42. *Jones*, 132 S. Ct. at 955–56 (quoting *Illinois v. Lidster*, 540 U.S. 419, 426 [2004]) (Sotomayor, J., concurring).

43. *Jones*, 132 S. Ct. at 956 (Sotomayor, J., concurring).

44. *Jones*, 132 S. Ct. at 963–964 (Alito, J., concurring).

45. Ibid. at 964 (Alito, J., concurring).

46. R. Jones, "Science and the Policeman," *Police Journal* 32:4 (October–December 1959): 236.

47. Andrew Tarantola, "The Big Daddy of Big Boom Disposal," *Gizmodo*, July 4, 2011, http://gizmodo.com/5816663/the-big-daddy-of-big-boom-disposal.

48. "San Francisco Police Bomb Robot Goes Haywire on Its Last Mission," *Chicago Tribune*, August 27, 1993, http://articles.chicagotribune.com/1993-08 -27/news/9308280017_1_bomb-squad-police-raid-squad-officers; "Robot Sent to Disarm Bomb Goes Wild in San Francisco," *New York Times*, August 28, 1993, http://www.nytimes.com/1993/08/28/us/robot-sent-to-disarm-bomb-goes -wild-in-san-francisco.html.

49. "Robot Sent to Disarm Bomb Goes Wild in San Francisco," *New York Times*.

50. Jonathan Strickland, "How Police Robots Work—Police Robot Tasks," *How Stuff Works*, http://science.howstuffworks.com/police-robot3.htm.

51. Erinn Cain, "Technological Advancements Give Law Enforcement Leg Up in Investigations," *MPNNow*, March 15, 2013, http://www.mpnnow.com/ topstories/x766880878/Technological-advancements-give-law-enforcement-leg -up-in-investigations?zc_p=1.

52. Jonathan Strickland, "How Police Robots Work," *How Stuff Works*, http://science.howstuffworks.com/police-robot.htm.

53. "Robot Team," *Washington County, Oregon Sheriff Department*, http:// www.co.washington.or.us/Sheriff/FightingCrime/SpecialResponseTeams/robot -team.cfm.

54. Chelsea J. Carter & Greg Botelho, "'CAPTURED!!!' Boston Police Announce Marathon Bombing Suspect in Custody," *CNN.com*, April 19, 2013, http://www.cnn.com/2013/04/19/us/boston-area-violence/index.html.

55. Noah Shachtman, "Armed Robots Pushed to Police," *Wired*, August 16, 2007, http://www.wired.com/dangerroom/2007/08/armed-robots-so/.

56. Jonathan Strickland, "How Police Robots Work—Police Robot Control," *How Stuff Works*, http://science.howstuffworks.com/police-robot1 .htm.

57. On the topic of *Robocop*, although astute readers might protest that *Robocop* actually predates all of the police robots discussed so far by appearing in theaters in 1987, three things should be noted. First, the movie took place at some point in the near future when Detroit was approaching financial ruin; so circa right now. Second, Robocop was not AI, he was a cyborg. Third, *Robocop* was fiction.

58. Jonathan Strickland, "How Police Robots Work."

59. J. D. Heyes, "Ohio Man Charged with Shooting Police Robot That Entered His Bedroom," *Natural News*, March 8, 2013, http://www.natural news.com/039402_police_robots_assault_Ohio.html.

60. In this section, I discuss domestic surveillance drones—aerial drones used within America for monitoring people, places, or things. In the next chapter, I discuss military drones, which include both attack drones and surveillance drones used by the military abroad.

61. Noah Shachtman, "Cops Demand Drones," *Wired*, August 10, 2007, http://www.wired.com/dangerroom/2007/08/cops-demand-dro/.

62. Anne Broache, "Police Agencies Push for Drone Sky Patrols," *Cnet.com*, August 9, 2007, http://news.cnet.com/Police-agencies-push-for-drone-sky-patrols/2100-11397_3-6201789.html.

63. "History," Federal Aviation Administration, https://www.faa.gov/about/history/brief_history/.

64. "Fact Sheet—Unmanned Aircraft Systems (UAS)," Federal Aviation Administration, press release, February 19, 2013, http://www.faa.gov/news/fact_sheets/news_story.cfm?newsId=14153.

65. "Unmanned Aircraft Operations in the National Airspace; Clarification of FAA Policy," Docket No. FAA-2006-25714 (February 6, 2007) (hereinafter referred to as Unmanned Aircraft Operations in the National Airspace; Clarification of FAA Policy).

66. Aviation Safety Unmanned Aircraft Program Office (Federal Aviation Administration), *Interim Operational Approval Guidance 08-01—Unmanned Aircraft Systems Operations in the U.S. National Airspace System* (March 13, 2008), 5–6 (hereinafter referred to as *Interim Operational Approval Guidance 08-01*).

67. Federal Aviation Administration, Form 7711-2, Application for Certificate of Waiver or Authorization, August 2008.

68. *Interim Operational Approval Guidance 08-01*, 6.

69. "Unmanned Aircraft (UAS)—Questions and Answers," Federal Aviation Administration, http://www.faa.gov/about/initiatives/uas/uas_faq/, last updated March 19, 2013.

70. "Unmanned Aircraft Operations in the National Airspace; Clarification of FAA Policy."

71. Federal Aviation Administration, "Model Aircraft Operating Standards," Department of Transportation Advisory Circular 91-57 (June 9, 1981).

72. "Fact Sheet—Unmanned Aircraft Systems (UAS)," Federal Aviation Administration, press release, February 19, 2013, http://www.faa.gov/news/fact_sheets/news_story.cfm?newsId=14153.

73. "Unmanned Aircraft Systems (UAS)," Federal Aviation Administration, http://www.faa.gov/about/initiatives/uas/, last updated on April 22, 2013.

74. "Unmanned Aircraft Systems (UAS)," Federal Aviation Administration; "List of Federal, State, or Local Agencies That Currently Hold or Have Held a Certificate of Authorization to Operate an Unmanned Aircraft System between November 2006 and June 30, 2011," Federal Aviation Administration, http://www.faa.gov/about/initiatives/uas/media/COA_Sponsor_List_042412.pdf.

75. Jennifer Lynch, "FAA Releases New Drone List—Is Your Town on the Map?" *Electronic Frontier Foundation*, February 7, 2013, https://www.eff.org/deeplinks/2013/02/faa-releases-new-list-drone-authorizations-your-local-law

-enforcement-agency-map; "List of FAA Drone Authorization List," *Electronic Frontier Foundation*, https://www.eff.org/sites/default/files/filenode/faa_coa_list-2012.pdf (released pursuant to Freedom of Information Act request).

76. Gerald L. Dillingham, *Unmanned Aircraft Systems—Continued Coordination, Operational Data, and Performance Standards Needed to Guide Research and Development,* GAO-13-346T, http://www.gao.gov/assets/660/652223.pdf (Washington, DC: U.S. Government Accountability Office, February 15, 2013) (Testimony before the Subcommittee on Oversight, Committee on Science, Space, and Technology, House of Representatives).

77. Brian Bennett and Joel Rubin, "Drones Are Taking to the Skies in the U.S.," *Los Angeles Times*, http://www.latimes.com/news/nationworld/nation/la-na-domestic-drones-20130216,0,3374671.story.

78. Jennifer Lynch, "Just How Many Drone Licenses Has the FAA Really Issued?" *Electronic Frontier Foundation*, February 21, 2013, https://www.eff.org/deeplinks/2013/02/just-how-many-drone-licenses-has-faa-really-issued.

79. "FAA Makes Progress with UAS Integration," Federal Aviation Administration, press release, May 14, 2012, http://www.faa.gov/news/updates/?newsId=68004.

80. "President Obama Signs FAA Reauthorization Bill into Law," *National Business Aviation Association*, February 15, 2012, http://www.nbaa.org/advocacy/issues/modernization/20120215-obama-signs-faa-reauthorization-bill-into-law.php.

81. Matthew L. Wald, "Current Laws May Offer Little Shield against Drones, Senators Are Told," *New York Times*, March 20, 2013, http://www.nytimes.com/2013/03/21/us/politics/senate-panel-weighs-privacy-concerns-over-use-of-drones.html?_r=0.

82. S. Smithson, "Drones over U.S. Get OK by Congress," *Washington Times*, February 7, 2012, http://www.washingtontimes.com/news/2012/feb/7/coming-to-a-sky-near-you/?page=all.

83. Catherine Crump and Jay Stanley, "Why Americans Are Saying No to Domestic Drones," *Slate*, February 11, 2013, http://www.slate.com/articles/technology/future_tense/2013/02/domestic_surveillance_drone_bans_are_sweeping_the_nation.html.

84. Aaron Cooper, "Drone Came within 200 Feet of Airliner Over New York," *CNN.com*, March 5, 2013, http://www.cnn.com/2013/03/04/us/new-york-drone-report/.

85. Florida Senate Bill 92 (Ch. 2013-33).

86. "Florida Police Want to Use Drones for Crowd Control," *ClickOrlando.com*, February 6, 2013, http://www.clickorlando.com/news/Florida-police-want-to-use-drones-for-crowd-control/-/1637132/18433078/-/cfswl2/-/index.html.

87. Crump and Stanley, "Why Americans Are Saying No to Domestic Drones."

88. "Virginia House of Delegates and Senate Approve Two Year Moratorium on Drones," American Civil Liberties Union, press release, February 6, 2013, http://www.aclu.org/criminal-law-reform/virginia-house-delegates-and-senate-approve-two-year-moratorium-drones. Virginia Governor Bob McDonnell subsequently added exceptions that would permit drone usage in certain situations.

89. W. J. Hennigan, "City in Virginia Passes Anti-Drone Resolution," *Los Angeles Times*, February 6, 2013, http://articles.latimes.com/2013/feb/06/business/la-fi-mo-drone-regulation-20130205.

90. "Seattle Mayor Ends Police Drone Efforts," Associated Press, February 7, 2013, http://www.usatoday.com/story/news/nation/2013/02/07/seattle-police-drone-efforts/1900785/.

91. Kara Kenney, "Bill to Crack Down on Drone Use in Indiana Dies in Committee," *TheIndyChannel.com*, February 25, 2013, http://www.theindychannel.com/news/local-news/bill-to-crack-down-on-drone-use-in-indiana-dies-in-committee.

92. "Making Connections at 45,000 Feet: Future UAVs May Fuel Up in Flight," press release, Defense Advanced Research Projects Agency, October 5, 2012, http://www.darpa.mil/NewsEvents/Releases/2012/10/05.aspx.

93. "What a UAV Can Do with Depth Perception," Press Release, Defense Advanced Research Projects Agency, December 6, 2012, http://www.darpa.mil/NewsEvents/Releases/2012/12/06.aspx.

94. Douglas Gantenbein, "Unmanned Traffic Jam," *Air & Space*, July 2009, http://www.airspacemag.com/flight-today/Unmanned-Traffic-Jam.html?c=y&story=fullstory.

95. Ben Coxworth, "A US$49 Personal Autonomous Micro UAV?" *Gizmag*, January, 28, 2013, http://www.gizmag.com/mecam-tiny-autonomous-uav/26007/.

96. "MeCam: Self Video Nano Copter to Point-and-Stream Yourself," Always Innovating, https://www.alwaysinnovating.com/products/mecam.htm.

97. Coxworth, "A US$49 Personal Autonomous Micro UAV?"

98. Air Traffic Organization NextGen & Operations Planning Office of Research and Technology Development (Federal Aviation Administration), *Unmanned Aircraft System Regulation Review*, DOT/FAA/AR-09/7 (Washington, DC: Federal Aviation Administration, September 2009), vii, 2–3.

99. See 49 USC § 40102, available at http://www.gpo.gov (accessed April 29, 2013); 14 CFR 1.1, available at http://www.gpo.gov (accessed April 29, 2013).

100. See U.S. Department of Transportation Federal Aviation Administration, Order 8130.34A, re: Airworthiness Certification of Unmanned Aircraft Systems and Optionally Piloted Aircraft (October 27, 2010); 14 CFR §§ 21.191, 21.193, 21.195, and 91.319, available at http://www.gpo.gov (accessed April 29, 2013).

101. Unmanned Aircraft Systems (UAS)," Federal Aviation Administration.

102. 2012 FAA Modernization and Reform Act, Pub. L. 112-95, February 12, 2012.

103. See *Jones*, 132 S. Ct. at 956 (Sotomayor, J., concurring) (noting that the Justice takes specific attributes of GPS technology into her consideration of reasonable expectation of privacy) and at 962 (Alito, J. concurring) (noting that technology can change privacy expectations under the Fourth Amendment).

Chapter 7

1. *The Paquete Habana,* 175 U.S. 677 (1900).

2. There is a distinction between public international law, which deals with the interaction of nations, and private international law, which deals with the international interactions of people and organizations. When I refer in this chapter to international law, I am referring to public international law.

3. United Nations, *Charter of the United Nations*, October 24, 1945, 50 Stat. 1031, http://www.refworld.org/docid/3ae6b3930.html, Art. 19.

4. See Mark W. Janis and John E. Noyes, *International Law Cases and Commentary*, 3rd ed. (St. Paul, MN: Thomson/West, 2006), 263.

5. See Frederic L. Kirgis, "United States Dues Arrearages in the United Nations and Possible Loss of Vote in the UN General Assembly," *ASIL Insights,* July 1998, http://www.asil.org/insigh21.cfm.

6. J. L. Bierly, *The Law of Nations,* 6th ed. (Oxford, UK: Clarendon Press, 1963), 5–6. This is hardly a unanimous opinion, but is useful in terms of an overview of international law. See Stéphane Beaulac, "The Westphalian Legal Orthodoxy—Myth or Reality?" *Journal of the History of International Law* 2: 2 (February 2000): 148–77.

7. Baron S. A. Korff, "An Introduction to the History of International Law," *American Journal of International Law* 18:2 (1924): 246–47.

8. 1 Maccabees 8:1-29 (*Good News Bible*); Mark W. Janis, "An Introduction to International Law," in Mark W. Janis and John E. Noyes, eds., *International Law Cases and Commentary,* 3rd ed. (St. Paul, MN: Thomson/West, 2006), 1; Arthur Nussbaum, "The Significance of Roman Law in the History of International Law," *University of Pennsylvania Law Review* 100:5 (March 1952): 679–80; "History of the International Court of Justice," International Court of Justice, http://www.icj-cij.org/court/index.php?p1=1&p2=1.

9. Nussbaum, 680–81.

10. Korff, 247.

11. Edward D. Re, "International Law and the United Nations," *St. John's Law Review* 21:1 (November 1946): 147–48.

12. See Nussbaum, 680–83.

13. Bierly, 5; Leo Gross, "The Peace of Westphalia 1648–1948." *American Journal of International Law* 42:1 (January 1948): 21–22.

14. Janis, "An Introduction to International Law," 2.

15. "History of the International Court of Justice."

16. See Janis and Noyes, 27.

17. *Statute of the International Court of Justice*, June 26, 1945, 59 Stat. 1055, http://www.icj-cij.org/documents/index.php?p1=4&p2=2&p3=0, Art. 38.

18. Janis and Noyes, 27–29.

19. Ibid., 29.

20. James Brown Scott, "The Codification of International Law," *American Journal of International Law* 18:2 (1924): 263–64.

21. Janis, "An Introduction to International Law," 2.

22. "History of the International Court of Justice."

23. Ibid.

24. David D. Caron, "War and International Adjudication: Reflections on the 1899 Peace Conference," *American Journal of International Law* 94:1 (January 2000): 15–22.

25. Janis, "An Introduction to International Law," 466–67.

26. "UN Welcomes South Sudan as 193 Member State," UN News Centre, press release. July 14, 2011, http://www.un.org/apps/news/story.asp?NewsID=39034.

27. Oscar Schacter, "United Nations Law," *American Journal of International Law* 88:1 (January 1994): 2–4.

28. "Nuclear Weapons," United Nations Office for Disarmament Affairs, http://www.un.org/disarmament/WMD/Nuclear/.

29. *Treaty on the Non-Proliferation of Nuclear Weapons*, March 5, 1970, 729 U.N.T.S. 161, http://disarmament.un.org/treaties/t/npt/text.

30. *Treaty Banning Nuclear Weapon Tests in the Atmosphere, in Outer Space and Under Water*, October 10, 1963, 14 U.S.T. 1313, http://www.un.org/disarmament/WMD/Nuclear/pdf/Partial_Ban_Treaty.pdf.

31. Statute of the International Atomic Energy Agency, July 29, 1957, 276 U.N.T.S. 3, http://www.iaea.org/About/statute.html, Arts. I & III.

32. "History." International Telecommunications Union. http://www.itu.int/en/about/Pages/history.aspx.

33. *Convention on the Prohibition of the Use, Stockpiling, Production and Transfer of Anti-Personnel Mines and on Their Destruction*, March 1, 1999, 2056 U.N.T.S 241, http://www.icbl.org/index.php/icbl/Treaty/MBT/Treaty-Text-in-Many-Languages/English (hereinafter referred to as the *Land Mine Convention*).

34. "Final Report," Meeting of the State Parties to the Convention on the Prohibition of the Use, Stockpiling, Production and Transfer of Anti-Personnel Mines and to Their Destruction, APLC/MSP.1/1999/1, Maputo, Mozambique (May 20, 1999), http://www.apminebanconvention.org/fileadmin/pdf/mbc/MSP/1MSP/1msp_final_report_en.pdf.

35. *Land Mine Convention*, Art. 6.

36. Anthony A. D'Amato, "The Concept of Special Custom in International Law," *American Journal of International Law* 63:2 (April 1969): 212.

37. Janis and Noyes, 92.

38. "States Parties to the Convention," *Land Mine Convention*, http://www.apminebanconvention.org/states-parties-to-the-convention/.

39. Mark Landler, "White House Is Being Pressed to Reverse Course and Join Land Mine Ban," *New York Times*, May 7, 2010, http://www.nytimes.com/2010/05/08/world/americas/08mine.html?_r=0.

40. Andrew C. S. Efaw, "The United States Refusal to Ban Landmines: The Intersection between Tactics, Strategy, Policy and International Law," *Military Law Review* 159 (1999): 100–101.

41. Having said that, it should be noted that the United States is a party to the Convention on Certain Conventional Weapons, which addresses land mines but permits their use.

42. Efaw, 94.

43. "Arguments for a Ban," *International Campaign to Ban Landmines*, http://www.icbl.org/index.php/icbl/Problem/Landmines/Arguments-for-a-Ban.

44. Efaw, 94.

45. Jill M. Sheldon, "Nuclear Weapons and the Laws of War: Does Customary International Law Prohibit the Use of Nuclear Weapons in All Circumstances?" *Fordham International Law Journal* 20:1 (November 1996): 182.

46. Brendan Gogarty and Isabel Robinson, "Unmanned Vehicles: A (Rebooted) History Background and Current State of the Art," *Journal of Law, Information and Science* 21:2 (2011/2012): 3, n. 5.

47. Charles Perley, 1863, Improvement in Discharging Explosive Shells from Balloons, U.S. Patent 37,771, issued February 24, 1863.

48. Gogarty and Robinson, 4, n. 7.

49. Ed Darack, "A Brief History of Unmanned Aircraft," *Air & Space*, May 18, 2011, http://www.airspacemag.com/multimedia/A-Brief-History-of-Unmanned-Aircraft.html?c=y&page=4&navigation=thumb#IMAGES.

50. John Sifton, "A Brief History of Drones," *The Nation*, February 27, 2012, http://www.thenation.com/article/166124/brief-history-drones#.

51. Darack, "A Brief History of Unmanned Aircraft," *Air & Space Magazine*, May 18, 2011, http://www.airspacemag.com/multimedia/A-Brief-History-of-Unmanned-Aircraft.html?c=y&page=8&navigation=thumb#IMAGES and http://www.airspacemag.com/multimedia/A-Brief-History-of-Unmanned-Aircraft.html?c=y&page=9&navigation=thumb#IMAGES.

52. Darack, http://www.airspacemag.com/multimedia/A-Brief-History-of-Unmanned-Aircraft.html?c=y&page=9&navigation=thumb#IMAGES.

53. Gogarty and Robinson, 4, n. 9.

54. Ibid., 6–7.

55. Paul Joseph Springer, *Military Robots and Drones* (Santa Barbara, CA: ABC-Clio, 2013), 22; Richard J. Newman, "The Little Predator That Could," *Air Force* magazine 85:3 (March 2002), http://www.airforce-magazine.com/MagazineArchive/Pages/2002/March%202002/0302predator.aspx.

56. Gogarty and Robinson, 9.

57. Springer, 22.

58. Newman.

59. Springer, 22.

60. Sifton, "A Brief History of Drones."

61. United States Government Accountability Office, "Unmanned Aircraft Systems: Improved Planning and Acquisition Strategies Can Help Address Operational Challenges," testimony before the U.S. House, Subcommittee on Tactical Air and Land Forces, April 6, 2006 (Washington., DC: Government Accountability Office, 2006), 5.

62. Charles Levinson, "Israeli Robots Remake Battlefield; Nation Forges Ahead in Deploying Unmanned Military Vehicles by Air, Sea, and Land," *Wall Street Journal*, January 13, 2010; "Predator Drones and Unmanned Aerial Vehicles (UAVs)," *New York Times*, http://topics.nytimes.com/top/reference/timestopics/subjects/u/unmanned_aerial_vehicles/index.html.

63. Gogarty and Robinson, 11–12.

64. Philip Alston, "Lethal Robotic Technologies: The Implications for Human Rights and International Humanitarian Law," *Journal of Law, Information and Science* 21:2 (2011/2012): 41.

65. Gogarty and Robinson, 12; see United States Airforce, *United States Unmanned Aircraft Systems Flight Plan 2009-2047*, May 18, 2009, http://www.fas.org/irp/program/collect/uas_2009.pdf, 25–27 (hereinafter referred to as *United States Unmanned Aircraft Systems Flight Plan*).

66. *United States Unmanned Aircraft Systems Flight Plan*, 25–26.

67. Ibid., 26–27.

68. Gogarty and Robinson, 13.

69. Newman.

70. *United States Unmanned Aircraft Systems Flight Plan*, 26–27; Gogarty and Robinson, 14.

71. Gogarty and Robinson, 14, n. 72.

72. Ibid., 2.

73. Philip Alston, *Interim Report of the Special Rapporteur of the Human Rights Council on Extrajudicial, Summary or Arbitrary Executions*, U.N. Document A/65/321, August 23, 2010, http://daccess-ods.un.org/TMP/563504.472374916.html, 13.

74. Ibid., 14.

75. Ibid., 15.

76. Alston, "Lethal Robotic Technologies," 44.

77. Cheryl Pellerin, "Robots Could Save Soldiers' Lives, Army General Says," *American Forces Press Service*, August 17, 2011, http://www.defense.gov/news/newsarticle.aspx?id=65064.

78. Peter Finn, "A Future for Drones: Automated Killing," *Washington Post*, September 19, 2011, http://articles.washingtonpost.com/2011-09-19/national/35273383_1_drones-human-target-military-base.

79. Department of Defense Directive 3000.09, November 21, 2012, available at http://www.dtic.mil/whs/directives/corres/pdf/300009p.pdf.

80. Chris Woods and Christina Lamb, "Obama Terror Drones: CIA Tactics in Pakistan Include Targeting Rescuers and Funerals," *The Bureau of Investigative Journalism.* February 4, 2012, http://www.thebureauinvestigates .com/2012/02/04/obama-terror-drones-cia-tactics-in-pakistan-include-targeting -rescuers-and-funerals/.

81. Peter W. Singer, "Attack of the Military Drones," *Brookings,* June 27, 2009, http://www.brookings.edu/research/opinions/2009/06/27-drones-singer.

82. Human Rights Council, *Interim Report of the Special Rapporteur of the Human Rights Council on Extrajudicial, Summary or Arbitrary Executions, Philip Alston—Addendum,* U.N. Document A/HRC/14/24/Add.6, May 28, 2010, http://unispal.un.org/UNISPAL.NSF/0/.

83. Alston, *Interim Report,* 24.

84. Laurie R. Blank, "After 'Top Gun': How Drone Strikes Impact the Law of War," *University of Pennsylvania Journal of International Law* 33:3 (Spring 2012): 681. In note 27 on this page, Blank provides a long list of key sources in international law governing the law of armed conflict, particularly the Geneva Conventions of August 14, 1949, and their protocols.

85. Blank, 681–83.

86. International Committee of the Red Cross, *Protocol Additional to the Geneva Conventions of 12 August 1949, and Relating to the Protection of Victims of International Armed Conflicts,* June 8, 1977, 1125 U.N.T.S. 3, Art. 51(4), http://www.refworld.org/docid/3ae6b36b4.html (hereinafter referred to as *Additional Protocol I*).

87. International Conferences (The Hague), *Hague Convention (IV) Respecting the Laws and Customs of War on Land and Its Annex: Regulations Concerning the Laws and Customs of War on Land,* October 18, 1907, http:// www.refworld.org/docid/4374cae64.html, Art. 23.

88. *Legality of the Threat or Use of Nuclear Weapons, Advisory Opinion.* 1996 I.C.J. 226, 257 (hereinafter referred to as Nuclear Weapons).

89. Nuclear Weapons, 275.

90. Blank, 686.

91. John Pike, "Coming to the Battlefield: Stone-Cold Robot Killers," *Washington Post,* January 4, 2009. http://articles.washingtonpost.com/2009 -01-04/opinions/36913238_1_robotic-aircraft-uavs-moore-s-law.

92. P. W. Singer, "Military Robots and the Laws of War," *The New Atlantis* 23 (Winter 2009), http://www.thenewatlantis.com/publications/military-robots -and-the-laws-of-war.

93. See Chris Jenks, "Law from Above: Unmanned Aerial Systems, Use of Force, and the Law of Armed Conflict," *Notre Dame Law Review* 85:3 (2009): 650–51.

94. Blank, 686–87.

95. Ibid., 689–90.

96. See *Declaration Renouncing the Use, in Time of War, of Explosive Projectiles Under 400 Grammes Weight,* December 11, 1868, http://www.icrc

.org/applic/ihl/ihl.nsf/Article.xsp?action=openDocument&documentId=568842C
2B90F4A29C12563CD0051547C, Preamble.

97. Blank, 694–98.

98. Alston, *Interim Report*, 24. Among the articles that rely on Alston's
analysis are: Aaron M. Drake, "Current U.S. Air Force Drone Operations and
Their Conduct in Compliance with International Humanitarian Law—An
Overview." *Denver Journal of International Law and Policy* 39:4 (Fall 2011):
629–60; Laurie R. Blank, "After 'Top Gun': How Drone Strikes Impact the
Law of War," *University of Pennsylvania Journal of International Law* 33:3
(Spring 2012): 675–718; Michael N. Schmitt, "Unmanned Combat Aircraft
Systems and International Humanitarian Law: Simplifying the Oft Benighted
Debate," *Boston University International Law Journal* 30:2 (Summer 2012):
595–620; among others.

99. Human Rights Watch, *Losing Humanity: The Case against Killer Robots*
(Human Rights Watch, 2012): 30–32.

100. See Department of Defense Directive 3000.09, Enclosures 3 and 4.

101. *Additional Protocol I*, Art. 57(1).

102. *Prosecutor v. Galić.* Case No. IT-98-29-T. Judgment and Opinion.
International Criminal Tribunal for the Former Yugoslavia (December 5,
2003), Para. 58.

103. Brendan Gogarty and Meredith Hagger, "The Laws of Man over
Vehicles Unmanned: The Legal Response to Robotic Revolution on Sea, Land
and Air," *Journal of Law, Information and Science* 19:1 (2008): 123.

104. Philip Alston, "Lethal Robotic Technologies: The Implications
for Human Rights and International Humanitarian Law," *Journal of Law,
Information and Science* 21:2 (2011/2012): 51.

105. Drake, 652–53.

106. "Drone Pilots May Need Distractions," *Discover TechNewsDaily*,
December 12, 2012, http://news.discovery.com/tech/robotics/drone-pilots
-distraction-121127.htm.

107. Bart Elias, *Pilotless Drones: Background and Considerations for
Congress Regarding Unmanned Aircraft Operations in the National Airspace
System* (Washington, DC: Congressional Research Service, September 10,
2012), http://www.fas.org/sgp/crs/natsec/R42718.pdf, 9.

108. Nicola Abé, "Dreams in Infrared: The Woes of an American Drone
Operator," *Spiegel Online*, December 14, 2012, http://www.spiegel.de/
international/world/pain-continues-after-war-for-american-drone-pilot-a-872726
.html.

109. *Land Mine Convention*, Art 1.

110. International Committee of the Red Cross, *The Mine Ban Convention:
Progress and Challenges in the Second Decade* (Geneva: International
Committee of the Red Cross, 2011), http://www.icrc.org/eng/assets/files/other/
icrc-002-0846.pdf, 2–5.

111. *Charter of the United Nations*, Art. 51: "Nothing in the present Charter shall impair the inherent right of individual or collective self-defence if an armed attack occurs against a Member of the United Nations".

112. *Convention on International Civil Aviation*, December 7, 1944, 15 U.N.T.S. 295, Art. 8, http://www.icao.int/publications/Documents/7300_cons.pdf.

113. *Geneva Convention on Road Traffic*, September 19, 1949, 125 U.N.T.S. 3, Arts. 8 & 10, http://treaties.un.org/doc/Treaties/1952/03/1952032 6%2003-36%20PM/Ch_XI_B_1_2_3.pdf.

Chapter 8

1. World Intellectual Property Organization (WIPO), *WIPO Intellectual Property Handbook*, 2nd ed. (Geneva: WIPO Publication, 2008): 17. Although WIPO actually tries to explain that patents are not monopolies, the explanation tends to support the idea that describing a patent as a monopoly is pretty spot-on.

2. See WIPO, 112.

3. See WIPO, 68.

4. WIPO, 40.

5. Ibid., 3.

6. Abraham S. Greenberg, "The Ancient Lineage of Trade-Marks," *Journal of the Patent Office Society* 33:12 (December 1951): 876.

7. Mladen Vukmir, "The Roots of Anglo-American Intellectual Property Law in Roman Law," *Journal of Law and Technology* 32 (1991–1992):127.

8. Greenberg, 879; Edward S. Rogers, "Some Historical Matter Concerning Trade Marks," *Michigan Law Review* 9:1 (November 1910): 30.

9. Ke Shao, "Look at My Sign!—Trademarks in China from Antiquity to the Early Modern Times," *Journal of the Patent and Trademark Office Society* 87:8 (August 2005): 654.

10. Greenberg, 877.

11. Vukmir, 131.

12. Greenberg, 879.

13. Vukmir, 130.

14. Harold C. Streibich, "The Moral Right of Ownership to Intellectual Property: Part I—From the Beginning to the Age of Printing," *Memphis State University Law Review* 6:1 (Fall 1975): 4–5 (hereinafter referred to as Streibich I); Vukmir, 132–33.

15. Streibich I, 24–29.

16. Ibid., 17 & 24–29.

17. See Streibich I, 17–18.

18. Streibich I, 35.

19. Harold C. Streibich, "The Moral Right of Ownership to Intellectual Property: Part II From the Age of Printing," to the Future," *Memphis State University Law Review* 7:1 (Fall 1976): 52–55 (hereinafter referred to as Streibich II).

20. Orit Fishchman-Afori, "The Evolution of Copyright Law and Inductive Speculations as to Its Future," *Journal of Intellectual Property Law* 19:2 (Spring 2012): 242–45.

21. See Streibich II, 55.

22. Fishchman-Afori, 245–47.

23. Vukmir, 130–31; see Streibich I, 4–10.

24. Vukmir, 134.

25. P. J. Federico, "Origin and Early History of Patents," *Journal of the Patent Office Society* 11:7 (July 1929): 292.

26. Susan Sell, "Intellectual Property and Public Policy in Historical Perspective: Contestation and Settlement," *Loyola of Los Angeles Law Review*, 38:1 (Fall 2004): 273; Federico, 293.

27. Federico, 293–94.

28. Fishchman-Afori, 249–50.

29. Ibid., 250.

30. *Paris Convention for the Protection of Industrial Property*, March 20, 1883, Arts. 2 &3, http://www.wipo.int/treaties/en/ip/paris/trtdocs_wo020.html.

31. *Paris Convention*, Art. 1.

32. *Berne Convention for the Protection of Literary and Artistic Works,* September 9, 1886, Art. 5, http://www.wipo.int/treaties/en/ip/berne/trtdocs _wo001.html.

33. *Berne Convention*, Art. 2.

34. WIPO, *Trademark Law Treaty*, October 27, 1994, 2037 U.N.T.S. 35, http://www.wipo.int/treaties/en/ip/tlt/trtdocs_wo027.html; WIPO, *Singapore Treaty on the Law of Trademarks*, March 27, 2006, *Singapore Treaty on the Law of Trademarks*, http://www.wipo.int/treaties/en/ip/singapore/singapore _treaty.html.

35. *Treaty on Intellectual Property in Respect of Integrated Circuits*, March 26, 1989, 28 I.L.M. 447 (1989), http://www.wipo.int/treaties/en/ip/ washington/trtdocs_wo011.html.

36. *International Convention for the Protection of Performers, Producers of Phonograms and Broadcasting Organizations*, October 26, 1961, 496 U.N.T.S. 43, http://treaties.un.org/doc/Treaties/1964/05/19640518%2002-04%20AM/Ch _XIV_3p.pdf; *Convention for the Protection of Producers of Phonograms against Unauthorized Duplication of Their Phonograms*, October 26, 1971, 866 U.N.T.S. 67, http://www.wipo.int/treaties/en/ip/phonograms/trtdocs_wo023.html.

37. See *Paris Convention*, Art. 3.

38. See *Berne Convention*, Art. 6bis (referring to an author's death) and Appendix, Art. IV (using the pronoun "he" to refer to an author, rather than it).

39. WIPO, *WIPO Copyright Treaty*, December 20, 1996, 36 I.L.M. 65 (1997), Arts. 4 & 5, http://www.wipo.int/treaties/en/ip/wct/trtdocs_wo033.html.

40. *Copyright Treaty*, Art. 8.

41. "Summary of WIPO Copyright Treaty (WCT)(1996)," WIPO, http:// www.wipo.int/treaties/en/ip/wct/summary_wct.html.

42. U.S. Constitution Art. I, Sec. 8, Cl. 8.

43. 17 USC §201(a), available at http://www.gpo.gov (accessed April 29, 2013).

44. 17 USC §302(a), available at http://www.gpo.gov (accessed April 29, 2013).

45. 17 USC §302(b), available at http://www.gpo.gov (accessed April 29, 2013).

46. Copyright Act of 1790, 1 Statutes At Large 124, available at http://www .copyright.gov/history/1790act.pdf.

47. Peter K. Yu, *Intellectual Property and Information Wealth: Copyright and Related Rights* (Westport, CT: Greenwood Publishing Group, 2007): 143.

48. "United States Copyright Office—A Brief Introduction and History," U.S. Copyright Office, http://www.copyright.gov/circs/circ1a.html.

49. Copyright Act of 1976, Pub. L. 94-553, October 19, 1976.

50. 35 USC §101, available at http://www.gpo.gov (accessed April 29, 2013).

51. Leahy-Smith America Invents Act, Pub L. 112-29, September 16, 2011. When it went into effect in 2012, the Leahy-Smith America Invents Act changed the underlying principle of the American patent system that has existed since the birth of the nation. Until this act, America operated under a "first to invent" system, meaning that even without a patent, if an inventor used an invention, he or she was still entitled to a patent. As of March 16, 2013, no one is entitled to a patent without filing with the U.S. Patent and Trademark Office first.

52. 35 USC §154(a)(2), available at http://www.gpo.gov, accessed 4-29-2013.

53. *Random House Dictionary*, s.v. "inventor."

54. Ibid., s.v. "author."

55. 35 USC §100(d), available at http://www.gpo.gov (accessed April 29, 2013).

56. 35 USC §100(f), available at http://www.gpo.gov (accessed April 29, 2013).

57. 17 USC §203(a), available at http://www.gpo.gov (accessed April 29, 2013).

58. 17 USC §101, available at http://www.gpo.gov (accessed April 29, 2013).

59. See 17 USC §302(c), available at http://www.gpo.gov (accessed April 29, 2013).

60. David Cope, *Tinman Too: A Life Explored* (Bloomington, IN: iUniverse, 2012), 297.

61. Cope, 297–98.

62. Ryan Blitstein, "Triumph of the Cyborg Composer," *Pacific Standard* , February 22, 2010, http://www.psmag.com/culture/triumph-of-the-cyborg-composer-8507/.

63. See Blitstein, "Triumph of the Cyborg Composer."

64. Tim Adams, "David Cope: 'You Pushed the Button and Out Came Hundreds and Thousands of Sonatas,'" *The Guardian*, July 10, 2010, http://www.guardian.co.uk/technology/2010/jul/11/david-cope-computer-composer.

65. Blitstein, "Triumph of the Cyborg Composer."

66. Adams, "David Cope: 'You Pushed the Button and Out Came Hundreds and Thousands of Sonatas.'"

67. Cope, 299.

68. Adams, "David Cope: 'You Pushed the Button and Out Came Hundreds and Thousands of Sonatas.'"

69. Blitstein, "Triumph of the Cyborg Composer."

70. David Cope, e-mail interview with author, April 7, 2013.

71. Blitstein, "Triumph of the Cyborg Composer."

72. Adams, "David Cope: 'You Pushed the Button and Out Came Hundreds and Thousands of Sonatas.'"

73. Blitstein, "Triumph of the Cyborg Composer."

74. Adams, "David Cope: 'You Pushed the Button and Out Came Hundreds and Thousands of Sonatas.'"

75. "Robopainter," *Studio360*, December 16, 2011, http://www.studio360 .org/2011/dec/16/robopainter/.

76. "Biography of Harold Cohen, Creator of AARON," Kurzweil Cyber Art Technologies, http://www.kurzweilcyberart.com/aaron/hi_cohenbio.html.

77. Harold Cohen, "The Further Exploits of Aaron, Painter," *Stanford Humanities Review* 4:2 (July 1995): 141–58 (hereinafter referred to as Cohen I).

78. "Ask the Scientists—Aaron the Artist—Harold Cohen," *PBS.com*, http:// www.pbs.org/safarchive/3_ask/archive/qna/3284_cohen.html.

79. "Robopainter," *Studio360*.

80. Harold Cohen, "How to Draw Three People in a Botanical Garden," *AAAI-88—Proceedings of the Seventh National Conference on Artificial Intelligence* (St. Paul, MN: AAAI, 1989), 847.

81. "Ask the Scientists—Aaron the Artist—Harold Cohen."

82. Cohen I, 158.

83. Mark K. Anderson, "'Aaron': Art from the Machine," *Wired*, May 12, 2001, http://www.wired.com/culture/lifestyle/news/2001/05/43685?current Page=all.

84. Cohen I, 158.

85. "'The Policeman's Beard' Was Largely Prefab!" *Robot Wisdom*, last updated September 1997, http://www.robotwisdom.com/ai/racterfaq.html.

86. "'The Policeman's Beard' Was Largely Prefab!" *Robot Wisdom*.

87. See "Racter," *Futility Closet*, August 15, 2012, http://www.futilitycloset. com/2012/08/15/racter/.

88. Joyce R. Slater, "'Just This Once' May Prove to Be an Ironic Prophesy," *Chicago Tribune*, July 29, 1993, http://articles.chicagotribune.com/1993-07-29/ features/9307290377_1_jacqueline-susann-dead-heat-french.

89. Peter Laufer, "Hack in the Valley of the Dolls," *Mother Jones*, May– June 1993, 15.

90. "Hal Is Back, and Writing Best-Sellers," *Baltimore Sun*, July 8, 1993, http://articles.baltimoresun.com/1993-07-08/news/1993189090_1_scott-french -computer-susann.

91. Steve Lohr, "In Case You Wondered, a Real Human Wrote This Column," *New York Times*, September 10, 2011, http://www.nytimes.com/2011/09/11/business/computer-generated-articles-are-gaining-traction.html?pagewanted=all&_r=0.

92. Taylor Bloom, "Automated Insights Using Big Data to Change How Fans Consume Sports," *Sport Techie*, March 6, 2013, http://www.sporttechie.com/2013/03/06/automated-insights-using-big-data-to-change-way-fans-consume-sports/.

93. Bloom, "Automated Insights Using Big Data to Change How Fans Consume Sports."

94. Patrick Seitz, "Narrative Science Turning Big Data into Plain English," *Investors.com*, August 21, 2012, http://news.investors.com/technology/082112-622940-narrative-science-takes-data-analytics-to-next-level.htm?p=full.

95. Buster Brown, "Robo-Journos Put Jobs in Jeopardy," *Huffington Post*, July 19, 2012, http://www.huffingtonpost.com/buster-brown/robo-journalism_b_1683564.html.

96. Steve Levy, "Can an Algorithm Write a Better News Story Than a Human Reporter?" *Wired*, April 24, 2012, http://www.wired.com/gadgetlab/2012/04/can-an-algorithm-write-a-better-news-story-than-a-human-reporter/all/1.

97. Brown, "Robo-Journos Put Jobs in Jeopardy;" Levy, "Can an Algorithm Write a Better News Story Than a Human Reporter?"

98. Farhad Manjoo, "Will Robots Steal Your Job? Part III Robottke," *Slate*, September 27, 2011, http://www.slate.com/articles/technology/robot_invasion/2011/09/will_robots_steal_your_job_4.single.html.

99. Bloom, "Automated Insights Using Big Data to Change How Fans Consume Sports."

100. Manjoo, "Will Robots Steal Your Job? Part III Robottke"; Levy, "Can an Algorithm Write a Better News Story Than a Human Reporter?"

101. Levy, "Can an Algorithm Write a Better News Story Than a Human Reporter?"

102. Lohr, "In Case You Wondered, a Real Human Wrote This Column."

103. David Cope, e-mail interview with author, April 10, 2013.

104. William T. Ralston, "Copyright in Computer-Composed Music: Hal Meets Handel," *Journal of the Copyright Society of the U.S.A.* 52:3 (Spring 2005): 294.

105. Ralston, 294, 300–301.

106. *Bleistein v. Donaldson Lithographing Co.*, 188 U.S. 239, 250 (1903).

107. *Burrow-Giles Lithographic Co. v. Sarony*, 111 U.S. 53, 57–58 (1884).

108. Ralston, 303.

109. David Cope, e-mail interview with author, April 7, 2013.

110. Adams, "David Cope: 'You Pushed the Button and Out Came Hundreds and Thousands of Sonatas.'"

111. "Ask the Scientists—Aaron the Artist—Harold Cohen."

112. Brown, "Robo-Journos Put Jobs in Jeopardy."

113. Ralston, 303.

114. U.S. Constitution Art. I, Sec. 8, Cl. 8.

115. Blitstein, "Triumph of the Cyborg Composer."

116. Ralph D. Clifford, "Intellectual Property in the Era of the Creative Computer Program: Will the True Creator Please Stand Up?" *Tulane Law Review* 71 (June 1997): 1695.

117. Ibid., 1698.

118. David Cope, e-mail interview with author, April 7, 2013.

119. Adams, "David Cope: 'You Pushed the Button and Out Came Hundreds and Thousands of Sonatas.'"

120. David Cope, e-mail interview with author, April 7, 2013.

121. Adams, "David Cope: 'You Pushed the Button and Out Came Hundreds and Thousands of Sonatas.'"

Chapter 9

1. "Farm Population Lowest Since 1850's," Associated Press, July 20, 1988, http://www.nytimes.com/1988/07/20/us/farm-population-lowest-since-1850-s .html.

2. "About Us," National Institute of Food and Agriculture, http://www .csrees.usda.gov/qlinks/extension.html.

3. Tyler Cowen, *The Great Stagnation* (New York: Dutton, 2011).

4. Seth Fletcher, "Yes, Robots Are Coming for Our Jobs—Now What?" *Scientific American*, March 8, 2013, http://www.scientificamerican.com/article .cfm?id=yes-robots-are-coming-for-our-jobs-now-what&page=1.

5. Gus Lubin, "Artificial Intelligence Took America's Jobs and It's Going to Take a Lot More," *Business Insider*, November 6, 2011, http://www.business insider.com/economist-luddites-robots-unemployment-2011-11.

6. Bernard Condon and Paul Wiseman, "Millions of Middle-Class Jobs Killed by Machines in Great Recession's Wake," *Huffington Post*, January 23, 2013, http://www.huffingtonpost.com/2013/01/23/middle-class-jobs-machines _n_2532639.html.

7. "Difference Engine: Luddite Legacy," *Science and Technology* (blog), *The Economist*, November 4, 2011, http://www.economist.com/blogs/babbage/2011/ 11/artificial-intelligence?fsrc=scn/tw/te/bl/ludditelegacy. Martin Ford in *The Lights in the Tunnel* indicates that AI could lead to the elimination of 60 percent of "average jobs."

8. Farhad Manjoo, "Will Robots Steal Your Job?" series of articles, *Slate*, September 26–30, 2011, http://www.slate.com/articles/technology/robot _invasion/2011/09/will_robots_steal_your_job.single.html.

9. "Difference Engine: Luddite Legacy."

10. See Fletcher, "Yes, Robots Are Coming for Our Jobs—Now What?"

11. Shaunacy Ferro, "Artificial Intelligence App Will Keep Tweeting as You after You Die," *PopSci*, March 11, 2013, http://www.popsci.com/technology/ article/2013-03/app-lets-you-tweet-great-beyond.

12. Anthony Berglas, *Artificial Intelligence Will Kill Our Grandchildren*, Draft 9, http://berglas.org/Articles/AIKillGrandchildren/AIKillGrandchildren .html#mozTocId818011. Also the plot of every *Terminator* movie.

13. There's the *Terminator* franchise. Also sounds a lot like a Keanu Reeves trilogy . . .

14. *The Matrix*. I knew that was in here somewhere.

15. U.S. Department of Transportation National Highway Traffic Safety Administration, "Traffic Safety Facts 2010 Data," DOT HS 811 630 (Washington, DC: U.S. Department of Transportation, June 2012), http://www .nhtsa.gov/staticfiles/nti/pdf/811630.pdf.

16. Matthew L. Wald, "Tougher Seat Belt Laws Save Lives, Study Finds," *New York Times*, November 17, 2003, http://www.nytimes.com/2003/11/17/ us/tougher-seat-belt-laws-save-lives-study-finds.html.

17. Martin DiCaro, "As Gov't Considers Regulations, Autonomous Car Boosters Show Off Plans," *Transportation Nation*, October 23, 2012, http:// transportationnation.org/2012/10/23/as-govt-considers-regulations-autonomous -cars-boosters-show-off-plans/.

18. "Look, No Hands," *The Economist*, September 1, 2012, http://www .economist.com/node/21560989.

19. William Saletan, "In Defense of Drones," *Slate*, February 19, 2013, http://www.slate.com/articles/health_and_science/human_nature/2013/02/ drones_war_and_civilian_casualties_how_unmanned_aircraft_reduce_collateral .html.

20. Noel Sharkey, "Don't Dismiss Robot Surgeons," *The Guardian*, August 26, 2008, http://www.guardian.co.uk/commentisfree/2008/aug/26/ health.robertwinston.

21. Tracey Schelmetic, "The Rise of the First Smart Cities," *ThomasNet.com Industry Market Trends Green & Clean Journal*, September 20, 2011, http:// news.thomasnet.com/green_clean/2011/09/20/the-rise-of-the-first-smart-cities/.

22. Nils J. Nilsson, "Artificial Intelligence, Employment and Income," *AI*, Summer 1984, 5–14.

23. Margaret A. Boden, "Artificial Intelligence as a Humanizing Force," *International Joint Conference on Artificial Intelligence* (Los Altos, CA: William Kaufman, 1983): 1197–98.

24. See Yvette Joy Liebesman, "The Wisdom of Legislating for Anticipated Technological Advancements," *John Marshall Review of Intellectual Property Law* 10:1 (2010):154–81.

25. 5 USC §553(b), available at http://www.gpo.gov (accessed April 29, 2013).

26. 5 USC §556, available at http://www.gpo.gov (accessed April 29, 2013).

27. 5 USC § 553(c), available at http://www.gpo.gov (accessed April 29, 2013).

28. Martin Shapiro, "A Golden Anniversary? The Administrative Procedures Act of 1946," *Regulation* 19:3 (1996): 3.

29. Nillson, "Artificial Intelligence, Employment and Income," 12.

30. Boden, 1197–98.

Index

About the Author

JOHN FRANK WEAVER is an attorney with McLane, Graf, Raulerson, and Middleton in Portsmouth, New Hampshire. As part of his practice, he speaks and publishes about the intersection of artificial intelligence and the law. He received his juris doctor degree from Boston College Law School and his bachelor's degree from Georgetown University. John lives in Portsmouth with his wife and daughter.

Follow him on Twitter: @RobotsRPeople.